The Sociology of Medical Regulation

John Martyn Chamberlain

The Sociology of Medical Regulation

An Introduction

Springer

John Martyn Chamberlain
Department of Social Science
University of Loughborough
Loughborough, Leicestershire, UK

ISBN 978-94-007-4895-8 ISBN 978-94-007-4896-5 (eBook)
DOI 10.1007/978-94-007-4896-5
Springer Dordrecht Heidelberg New York London

Library of Congress Control Number: 2012947932

Printed on acid-free paper

Springer is part of Springer Science+Business Media (www.springer.com)

Preface

This book provides an introduction to the sociological study of medical regulation. The governance of medical practitioners is dominated by the dilemma of how best to regulate the medical profession to ensure the public are protected from underperforming doctors while at the same time acknowledging that medical work can, and often does, have unintended outcomes. Traditionally, medical practitioners have been left to manage their own training and disciplinary arrangements on the basis that they possess high ethical standards in addition to their esoteric expertise, so they can be trusted to place their client's needs above their own. However, over the last three decades, the state has intervened in the field medical regulation and required medical practitioners adopt a more open, transparent and publicly accountable regulatory system which possesses more formal mechanisms for peer surveillance, appraisal and control. For example, in the United Kingdom, the 2008 Health and Social Care Act introduced a periodic review of a doctor's continued fitness to practise, called revalidation, that will be implemented nationally from 2012 onwards. The 2008 Act also reformed the medical professions key regulatory institution in the United Kingdom, the General Medical Council, and how complaints against doctors are handled. Such developments support the argument that we are now sitting on the cusp of far-reaching reform in medical regulation. It is therefore important to establish the current state of the art in the sociological study of the medical profession and its regulation, a state of affairs that led me to write this book.

In writing this book, my aim is to provide the reader with a definitive text that offers an up-to-date and comprehensive examination of the complex issues surrounding the regulation of the medical profession, but which is nevertheless written in an accessible way for both undergraduate and postgraduate students, be they from a social science or health-care profession background. The text aims to offer the reader an insight into key sociological theories surrounding medical regulation; a historically situated analysis of the contemporary relationship between medicine, the state and the public; an overview of relevant social scientific research; insight into possible future directions for medical governance; as well as end-of-chapter self-study tasks to consolidate chapter content.

The organisation of the chapters is as follows:

Chapter 1 provides an introduction to the book and its key themes. It outlines the important role medicine as a profession plays in the governance and social ordering of contemporary society and how the study of medical regulation must therefore bear in mind the broader sociopolitical and cultural context in which medical regulation operates. In doing so, the chapter sets the scene for the subsequent discussion of the historical development of modern medicine in Chap. 2.

Chapter 2 gives a historical account of the emergence of modern medicine against the background of how the humoral medical tradition, which dates back to the ancient Greeks, was gradually replaced from the eighteenth century onwards by a biomedical model of illness and disease. In telling the story of the subsequent development of biomedicine, the chapter discusses how the fledging medical profession utilised this approach to the treatment of illness and disease to establish a monopoly over the provision of health care from the mid-nineteenth century onwards.

Chapter 3 takes up the story of the development of the modern medical profession and brings it up to date by outlining the institutionalisation of the principle of professional self-regulation during the nineteenth century and exploring subsequent policy initiatives throughout the twentieth century, up to and including the 2008 Health and Social Care Act. This chapter provides the basis from which subsequent chapters explore the sociological study of medical regulation.

Chapter 4 outlines the sociological study of the medical profession from the early functionalist to later neo-Weberian, neo-Marxist and feminist viewpoints. It highlights how these latter approaches are critical of the tendency of the early functionalist tradition to unquestionably accept that the medical profession is a force for good within modern society. In doing so, the chapter outlines how the principle of professional self-regulation was increasingly criticised from the 1970s onwards for not protecting the public from instances of deliberate medical malpractice as well as underperforming doctors.

Chapter 5 takes up the discussion begun in Chap. 4 and outlines how sociologists have sought to answer the question of if medical autonomy is now in decline. It does this against the background of exploring the Foucauldian governmentality viewpoint. This argues that instead of being in decline medical autonomy is in fact being transformed into a new operational form as a result of the re-emergence of liberalism as an economic and political philosophy in Western nation-states from the 1980s onwards. The chapter concludes by synthesising the governmentality viewpoint with the neo-Weberian and feminist perspectives to argue that instead of being in decline medicine is undergoing a process of restratification whereby the profession increasingly splits into elite and rank and file segments and the former subject the latter to new and more intrusive forms of peer surveillance and control in order to maintain regulatory privileges (albeit in a more publicly accountable form).

Chapter 6 discusses the implementation of revalidation in the United Kingdom in comparison to similar developments internationally. Revalidation periodically retests the continued fitness to practise of a qualified doctor after they have completed their medical training. The chapter explores the implications of the introduction of revalidation for the restratification thesis.

Chapter 7 examines the handling of complaints against doctors and how fitness to practice cases proceed through the regulatory system, in order to identify if more doctors are being subject to disciplinary action as a result of recent reforms. In doing so, the chapter provides empirical support for the restratification thesis while at the same time acknowledging its limitations.

Chapter 8 concludes the book by discussing the need for further empirical research into medical regulation from the perspective of both doctors and patients. It also highlights the need for sociologists to remain aware of the broader social changes which are occurring in contemporary society and arguably shaping the regulatory arrangements for doctors.

I would like to take this opportunity to thank my students for showing some interest in this topic and inspiring me to write this book, alongside my publisher, Springer, for agreeing to support the project. I would also like to mention my daughter, Freyja, for without her none of this would be worthwhile. Finally, and if for no reason other than the fact that every girl should have a book dedicated to her once in her life, I would like to thank my ex-wife, Ellen, for continuing to be a good friend to me even though we no longer share a life together.

Contents

Chapter 1
Introduction: Medical Governance in the Risky Age of the Surveillance Society

Abstract This chapter outlines the focus of this book on contemporary developments in the regulation of the medical profession. In doing so, it notes that one of its key aims is to reinforce the need to recognise how current reforms in medical governance are bound up with broader changes in the nature of governing regimes within western nation-states as they seek to maintain the legitimacy of neo-liberal forms of governmentality. In doing so, this chapter outlines the main contents of subsequent chapters against the background of exploring how risk-management techniques and surveillance technologies have become dispersed throughout contemporary society, not just medicine's self-governing institutional mechanisms. To achieve this goal, this chapter outlines the development of contemporary writings on risk and how this sees surveillance as a necessary disciplinary mechanism by which political elites exercise legitimate social control over the population at large. In exploring this literature, this chapter concludes by noting that sociologists interested in the study of medical regulation need to move away from trying to identify if medical autonomy is in decline. Rather, this chapter sets the scene for the subsequent contents of this book. It does this by discussing how sociologists need to explore how current reforms in medical governance are bound up with a more general shift in how good governance is perceived and enacted by governing agencies under the risk-saturated social conditions associated with high modernity. End-of-chapter self-study tasks are provided so the reader can engage in further study in relation to chapter content.

Introduction

This book is concerned with the sociological analysis of the medical profession and how it is regulated. It outlines and discusses recent developments in medical regulation through a critical academic lens. However, in writing this book, I also seek to challenge a key assumption some sociologists start from when conducting an analysis of how medicine as a profession is (and should be) regulated. As will be discussed,

J.M. Chamberlain, *The Sociology of Medical Regulation: An Introduction*,
DOI 10.1007/978-94-007-4896-5_1, © Springer Science+Business Media Dordrecht 2013

historically medical regulation has operated within most western nation-states under what can be termed the ‘doctor knows best’ self-regulatory model. That is, given the esoteric and highly specialised nature of medical expertise, it has generally been though preferable for doctors to be left alone to collectively control professional training programmes and the standards by which an individual doctor’s professional competence (both technical and ethical) is judged. Indeed, historically, although the level of freedom from state, managerial-bureaucratic and consumer control has varied somewhat from country to country and culture to culture, the general modus operandi of medical regulation in the West for the last 200 years has been one of medical control via some form of collectively organised autonomous professional association, such as the royal medical colleges in the United Kingdom.

This state of affairs has led to sociologists, amongst others, to conclude that doctors possess medical autonomy, both individually and collectively. Of course, it is important to remember that the term autonomy is applied here in a relative manner through relating it to the level of control over working practices and the standards governing them possessed by different forms or types of workers. For it is certainly arguable that doctors possess autonomy over their working practices when they are compared to other occupational groups. Indeed, an entire field of study entitled the sociology of the professions has been built around the demarcation lines which surround occupational groups such as doctors that possess a large degree of autonomy over their working practices. This is because they possess a highly specialised and esoteric form of expertise which cannot be easily routinised into a series of step-by-step procedures that just anybody can follow to achieve certain desired ends. However, more lately, sociologists (along with certain patients’ right groups, politicians, media outlets, and indeed society at large) have argued that doctors need to be more open and transparent in how they manage their affairs. Arguments have been put forward that since it is possible for outsiders to judge some aspects of medical work, particularly in relation to ethical matters of probity, then medical regulation should be a more inclusive affair.

It is certainly the case that a series of high-profile medical malpractice cases and cover-ups have affirmed the need to reform how medicine as a profession is regulated. As a result, over the last two decades or so, there has been a gradual shift in how medical governance is practiced. This, in turn, has led some commentators to celebrate a successful challenge to traditional notions of medical privilege and a decline in medical autonomy as medical practitioners becoming increasingly subject to patient-consumerist and managerial-bureaucratic forms of surveillance and control. Providing the reader with an account of the historical development of this state of affairs is a key aim of this book. Indeed, Chaps. 2, 3, 4, and 5 are concerned with, firstly, outlining the emergence of medical power and autonomy as a result of growth of the biomedical worldview; secondly, tracing its subsequent alleged gradual decline through the challenge of unfolding developments in medical technology, alongside the contemporary growth of consumerist ideals and managerialist practices; and thirdly, charting how the response of medical elites to this state of affairs has reinforced that medical autonomy may well have been challenged but nevertheless does seem to be undergoing a process of transformation rather than outright decline. During this discussion, the idea is outlined that medicine has undergone a process of restratification into more pronounced elites and rank and file groups, with the former increasingly subjecting the latter to surveillance and control just as they themselves in turn become subject to performance appraisal mechanisms via state-appointed agencies, that is, National Health Service hospital management.

It is important for theoretical considerations to meet the test of empirical evidence. Consequently, Chaps. 6 and 7 are concerned with the critical appraisal of the restratification thesis in relation to the introduction of revalidation in the United Kingdom alongside recent trends in the hearing of fitness to practise cases. Yet the purpose of these chapters is not solely to provide evidence for the restratification thesis. Rather, they also point towards its inherent limitations, and in doing so they reinforce how as contemporary events in medical regulation unfold it is becoming increasingly necessary to adopt a more contextually aware analytical approach to the study of medical governance. This key point is taken up in Chap. 8. Indeed, it is this broader approach which I feel challenges current orthodoxy in the sociological study of medical governance. Bear with me for a moment while I try to place what I mean by this within the context of the aims of this book.

My starting point for writing this book is the urgent need to provide a thorough introduction to the sociological study of medical regulation. To this end, I trace much well-known ground without challenging too much the conceptual territory and orthodoxy surrounding the topic. This is so the reader who is unfamiliar with this field of study will find the following contents a useful general historical overview to the study of medical regulation and how it has been approached by sociologists. Yet, as will become clear, the sociological study of medical regulation is informed by two overlapping sub-disciplines: the sociology of the professions and the sociology of health and illness. Here, I think it is important to note that it is precisely the eclectic nature of the academic literature that underpins the sociological study of medicine which can lead many students of the field to begin their analysis of contemporary developments in medical regulation from the wrong analytical starting point. In essence, they do this when asking themselves, in one way or another, the question 'is medical autonomy in decline?' As will be discussed, the question of if medical autonomy is in decline or not is inherently problematic as it tends to ignore the fact that the operation of medicine (at both the day-to-day and more abstract levels) is very much bound up with the broader governance of society at large: In a very real sense to the patient seeking health care, the medical profession *is* the state. This point will be discussed in detail in subsequent chapters. Indeed, part of my reason for writing this book is that I want to highlight the importance of retaining sight of broader social changes occurring within western nation-states and how these have impacted on the governance of not just the medical profession, but social life as a whole.

This tradition of placing the analysis of medicine within a broader context possesses a long history within the sociological study of the professions and the broader sociology of health and illness literature, yet somewhat interestingly it is often underplayed or ignored within the subfield of analysis known as professional regulation. It is my intention in this book to attempt to highlight its continued relevance.

We will discuss in Chap. 4 how Marxist-oriented commentators during the 1970s and 1980s argued for the need to place the study of medicine within the context of the study of capital. Similarly, Foucault (1991) recognises the importance of maintaining the position that medical governance is a reflection of state governance. To my mind, the value of the Foucauldian governmentality viewpoint, which is outlined in some detail in Chap. 5, is that it reminds us how over the last three decades there has been a general transformation in the conditions under which good governance can be practiced in western nation-states and that recent developments in medical regulation are very much bound up with this transformation.

This transformation in how contemporary societies are governed goes well beyond the question of whether medical autonomy is in decline or not. Indeed, it is important to remember that reforms in medical governance are as much aimed at the object of governmentality – the population – as they are the medical profession. As will be discussed in some detail in subsequent chapters, neo-liberal forms of governmentality – which arguably dominate today's technologically advanced western nation-states – require tighter state control of professional groups, alongside the introduction of transparent performance appraisal mechanisms to manage their activities, as part of a broader governmental project to enable a more dispersed, citizen-responsibilised form of self-surveillance and performance management to take hold within society at large. Over the last three decades or so, the state has sought to expand its control over the population through somewhat ironically promoting, on one hand, greater freedom of choice, consumerist ideology and participatory forms of democracy, and, on the other hand, a more punitive form of welfarism which seeks to identify, survey, risk-manage and control those individuals who fail to meet the criteria for what constitutes good citizenship, that is, anybody who is not a healthy, law-abiding and productive worker-consumer. This state of affairs can be perhaps most clearly seen in the fact that the Anglo-American prison population has risen dramatically over the last three decades as neo-liberalism has emerged to dominant economic and political discourse on both sides of the Atlantic. Yet it is also important to recognise that it can also be seen in contemporary developments in medical regulation.

It is my contention, therefore, that the study of medical regulation must bear in mind this broader state of affairs and draw on forms of sociological and even criminological academic discourse concerned with core analytical concepts, such of risk, surveillance and governance. Indeed, if we are to fully account for nature of contemporary shifts in medical governance, then I would argue it is vitally important to examine the emergence of the risk-saturated conditions associated with the surveillance society and the role professional forms of expertise play within this. This should not be too difficult a task given the eclectic nature of the sociological study of professional forms of regulation: It is used to drawing on a range of differing perspectives to examine its subject. Yet the reader who is approaching the topic for the first time could perhaps be forgiven for wanting a clear starting point from which to begin. Hence, in the rest of this chapter, I will provide a broad introductory background to the emergence of risk and surveillance as organising background concepts to begin to explore the analysis of contemporary trends in medical regulation.

This discussion can be used by the reader as a point of reference to explore the contents of subsequent chapters. Furthermore, it is to this background that we will return in Chap. 8 when we conclude our exploration of contemporary trends in medical governance.

From Modernity to High Modernity

Although the art of healing is as old as human civilisation itself, it was not until the re-emergence of the scientific method under the banner of enlightenment idealism from the eighteenth century onwards that the distinctive form of modern medicine practice known as biomedicine came into being. The broad intellectual movement known as the enlightenment involved a complex array of social, economic, political, technological and cultural processes. It produced a very different form of social life than had existed in Europe up until the seventeenth century. Indeed, it completely changed how citizens of European nation-states viewed their place in the world. For example, for the first time, we see the growth of democratic ideals and notions surrounding social equality and opportunity. So we start to see the gradual rejection of traditional social hierarchy's which were built around the social structure present within the serf/lord relationship which had dominated much of the social world until this time. This is why this period in human history is sometimes referred to by social scientists such as Polanyi (1967) as the Great Transformation. We certainly are talking here about a complex and far-reaching form of social change. But for the sake of clarity, we can break down the birth of modernity into three key interrelated processes.

First is the enlightenment. What we have here is the re-emergence of the idea from classical Greece that the scientific method can be a valid source of human knowledge and a concurrent belief in the ability of reason and practical experiment to describe, explain and change the world around us. The idea of progress is bound up with this. Prior to this time, people by and large lived their lives out in a set cyclical way, very much in tune with the natural seasons of the world. But with the re-emergence of science, we see the growth of the idea that human beings can change the world around them and indeed control it to their satisfaction. This directly challenged traditional orthodoxies of the God-given order of the natural and social worlds, with monarchs at the top and serfs at the bottom. So we start to see sustained attempts at mass social transformation, such as the French Revolution or the American War of independence. Secondly, and related to the scientific progress bound up with the enlightenment, is a rapid process of urbanisation and industrialisation. Here, we see how gradual technological advances led to what was called the industrial revolution, with work opportunities increasingly shifting from rural farming areas to urban factory's as a result of scientific progress enabling the mass production of goods. Here, of course, we also start to see urban overcrowding, and so problems, such as poverty and disease, led to the early development of public health medicine. Thirdly and finally, tied up with the scientific progress of the enlightenment, social reform and rapid industrialisation, is the growth of capitalism and free market economics. In short, we see the beginnings of liberalism as a particular economic and political philosophy that believes in the need to set limits on the role of the state in the governance of society.

Liberalism is discussed in greater detail in Chaps. 4 and 5. For the moment, it is enough to say that classical liberalism is a critique of state reason which seeks to set limits on state power (Peters 2001). But not only does liberalism have a particular view of the role of the state, it also has a particular view of the nature of the individual. The concept of possessive individualism lies at the heart of classical liberalism (Macpherson 1962). This first emerged in the seventeenth and eighteenth centuries, through the works of a variety of writers, such as Thomas Hobbes, John Stuart Mills, Adam Smith, Thomas Locke, Jeremy Bentham and Herbert Spencer. Macpherson (1962) argues that for these thinkers the individual and their capabilities prefigure the circumstance into which she is born. In short, an individual's talents and who they are owe nothing to society, rather they own themselves, and as such are morally and legally responsible for themselves alone. In this viewpoint, the individual is naturally self-reliant and free from dependence on others. They need only enter into relationships with others because they help them pursue their self-interests. Bound up with this is viewpoint is the belief that society is a series of market-based relations made between self-interested subjects who are actively pursuing their own interests. It was argued that only by recognising and supporting this position politically and economically will the greatest happiness for the greatest number be achieved.

A key problem with this view of the individual is that it tends to ignore or underplay the value of existing social structure and inequalities therein. Indeed, a very real problem here is that individual members of society do not start their lives equally. This fact led social reformers in the nineteenth and twentieth centuries to advocate changes in working conditions, poor relief and public health. A huge literature was produced by social activists of the time, such as Henry Mayhem, linking inequality and poverty to disease and death (White 2001). Furthermore, contra the ethos of liberalism, after the financial crash of the 1920s, economists such as John Maynard Keynes tended to argue for a strong interventionist role for the state in regulating the market, protecting working and living conditions, as well as promoting public health. Adopting Keynesian economics to control the tendency of capitalism to operate in boom and bust cycles formed an important part of the foundation of the post-Second World War welfare state in the United Kingdom. However, as the twentieth century progressed sociologists began to notice that significant technological and sociocultural changes were occurring in the makeup of states, particularly from the 1960s onwards, while the rate of change rapidly intensified as the 1980s progressed. The birth of the personal computer and the rise of the mobile phone reinforced that the way people were living there lives and relating to one another and the world around them was gradually being fundamentally reorganised. There also seemed to be a distinctive shift towards more diverse, pluralistic and multicultural populations within western nation-states as a result of a rise in immigration due to the increasing availability of cheap international transport and advances in communication technologies. Such considerations led some to conclude that the processes of modernity had intensified and led to a period in human history characterised by an intense feeling of personal uncertainty to such a degree that we had entered the age of high modernity. For some, this signalled the emergence of the risk society.

The Rise of the Risk Society

For many social scientists, the re-emergence of liberalism from the 1980s coincided with a general social shift towards the conditions of late or high modernity. We certainly live in an increasingly interconnected, technologically advanced, globalised world where events and happenings occurring on the other side of the globe are immediately available for personal consumption (and arguably therefore immediately impact on the sociocultural and economic-political spheres). For social theorists such as Beck (1992) and Giddens (1990, 1991, 1999), a key defining feature of modern society – or late or high modernity as they call it – is that there has been *'a social impetus towards individualisation of unprecedented scale and dynamism…[which]…forces people – for the sake of their survival – to make themselves the centre of their own life plans and conduct'* (Beck and Beck-Gernsheim 2002: 31). Both Beck and Giddens argue that as capitalist-industrial society gives way under the tripartite forces of technology, consumerism and globalisation, there is a categorical shift in the nature of social structures and, more importantly, the relationship between the individual and society. Here, key sociological categories which have traditionally structured society increasingly lose their meaning. Hence, social categories such as race, gender and class, for example, increasingly no longer serve to restrict a person's social opportunities or define who they are as individuals to the extent they once did. Furthermore, as working conditions change, and the technology and com munication revolutions continue at pace, more than ever before individuals are required to make life-changing decisions concerning education, work, self-identity and personal relationships, in a world where traditional beliefs about social class, gender and the family are being overturned.

Now for many social theorists, this state of affairs has led to a concern with dangerousness and risk entering centre stage within society's institutional governing apparatus, alongside individual subject-citizen's personal decision-making process (Mythen 2004). One of the key risk theorists, Giddens (1990), talks about two forms of risk: external and manufactured risk. Put simply, external risks are those posed by the world around us and manufactured risks are created by human beings themselves. In essence, as Giddens explains, it is the difference between worrying about what nature can do with us – in the form of floods, famine and so on – and worrying about what we have done to the natural world via how we organise social life. But of course it is not that simple. Risk theorists argue that throughout human history societies have always sought to risk-manage threats, hazards and dangers. But these management activities have by and large been concerned with natural external risks, such as infectious diseases and famine.

However, in today's technologically advanced society, individuals are seen to be both the producers and minimisers of manufactured risk (Giddens 1990). That is, within the conditions of high modernity, risks are seen to be solely the result of human activity (Mythen 2004). Hence, manufactured risk takes over. Even events previously held to be natural disasters, such as floods and famine, are now held to be avoidable consequences of human activities that must be risk-managed (Lupton 2011). Hence, society's governing institutions and expert bodies need to become ever more collectively self-aware of their role in the creation and management of risk (Beck and Beck-Gernsheim 2002). For the individual, meanwhile, uncertainties now litter their pathway through life to such an extent that it appears to be loaded with real and potential risks. So they must seek out and engage with on a seemingly ever-growing number of information resources, provided by a myriad of sources, as they navigate through their world. In the risk society, *'[we] find more and more guidebooks and practical manuals to do with health, diet, appearance, exercise, lovemaking and many other things'* (Giddens 1991: 218).

Of course, this state of affairs all links in with the possessive liberalism view of the individual being responsible for themselves, and indeed, risk theorists such as Giddens and Beck talk about how we can see that since the 1960s and 1970s there has been a growing cultural and political discourse of rights and responsibilities emerging which seeks to regulate the individual while also arguing for the need for greater personal freedom. This leads us into another key feature of high modernity which is arguably central to the study of medical self-regulation. Namely, that within the risk society, a sense of growing (perhaps even mutual) distrust characterises the relationship between the public and experts (Giddens 1999). At the same time, a pervasive and seemingly increasingly necessary reliance on an ever-growing number of experts appears to be a key feature of the individuals' personal experience of everyday life (Mythen 2004). Interestingly, it was argued that this established the conditions for the public to challenge elitism and expert forms of knowledge. For under such changing social conditions, expert authority can no longer simply stand on the traditional basis of position and status. Not least of all because an individuals' growing need to manage risk and problem solve their everyday life, to make choices about who they are and what they should do, means that personal access to the technical and expert knowledge of the elite becomes more urgent than ever before, while the development of mass information sharing tools, such as the mobile phone, personal computer and the Internet, meant that knowledge and expertise is no longer the sole preserve of those elite few who have undergone specialist training. As Giddens (1991: 144–146) notes, *'technical knowledge is continually re-appropriated by lay agents...Modern life is a complex affair and there are many 'filter back' processes whereby technical knowledge, in one shape or another, is re-appropriated by lay persons and routinely applied in the course of their day-today activities...Processes of re-appropriation relate to all aspects of social life – for example, medical treatments, child rearing or sexual pleasure'*.

There is then a tension between experts and citizens, between those in power and those who are not, and this can perhaps most clearly be seen in relation to modern technological advancements, particularly in relation to the rise of surveillance technology. After all, surveillance is essential to the task of identifying and managing and controlling risk. Not least of all because under the neo-liberal social conditions associated with high modernity, it is by their ability to successfully manage risk that state legitimises its governing activities. Bound up with this, as we shall now turn to discuss, is the need for law-abiding citizens to allow the surveillance of their everyday life to become a normalised feature of everyday existence.

The Risk Society as the Surveillance Society

One of the key academics who have looked at the growth of surveillance in modern society is Lyon (1994, 200). Lyon (2001:33) defines the surveillance society as *'a situation in which disembodied surveillance has become societally pervasive'*. He goes on to say that the *'precise details of our personal lives are collected, stored, retrieved and processed every day within huge computer databases belonging to big corporations and government departments. This is the "surveillance society"'* (Lyon 1993:3). Lyon argues that they are two faces to surveillance: care and control. That is, on one hand, the growth of surveillance technology enables us to more successfully manage risk and so care for ourselves and our social groups, but on the other hand, it also enables governments and state agencies to monitor and control populations, particularly for signs of potential threat and risk. Hence, social sorting is a central feature of the application of surveillance technology at all levels. For surveillance possesses a classificatory imperative related to the ability to socially sort activities, peoples and events. This renders it a medium of power to control and risk-manage citizens so their behaviour reflects dominant social, cultural and commonsense norms and values. Furthermore, for risk theorists, it is no coincidence that western states such as America and the United Kingdom suffer from endemic forms of surveillance. This is because, as Lyon notes, the rise of the surveillance society may be traced to modernity's impetus to coordinate or control. Surveillance technology certainly is rooted in modernity and the rise of the modern nation-state and a concurrent increase in bureaucratic institutional organisation. For Giddens (1990:321), surveillance is bound up with the growth of modernity, for *'surveillance is fundamental to social organization of all types, the state being historically the most consequential form of organization, but nevertheless being only one organization among many others'*.

The work of Foucault (1979) has been very instructive in helping social scientists trace the historical development of the growth of surveillance. Foucault analysed historical penal documents from which he tied the development of modern surveillance to the punishment practices established in modern European prisons from the early eighteenth century onwards. His analysis reveals that during this period punishment shifted from the public spectacles of torture and execution, what Foucault calls exercises of monarchical power, to the techniques of what he terms 'soul training', which were mastered in the new prison regimes emerging at the time under the aegis of the enlightenment revolution. These, he holds, were geared towards the production of obedient and docile individuals, who in line with the enlightenment pursuit of engendering positive social change through the application of reason, were held to be redeemable, if handled correctly, and hence ultimately able to contribute to society in some small way, that is, if they are punished correctly with a view to changing their behaviour. Foucault termed this process carceral punishment. In that, it heralded a constant surveillance of inmates under a new kind of power, namely, disciplinary power, which over time he argued was gradually replacing monarchical power.

Foucault argues that as the nineteenth century progressed the prisons of Europe bore witness to the development a form of spatial and temporal control over prisoners via hierarchies of classification and control which sought to discipline both body and mind. Amongst other things, Foucault uses the daily prison timetable of routines to illustrate his points: with each prisoner's daily routine following a set series of disciplinary practices, ranging from washing to eating, to working and exercising, from the time they woke up in the morning until the time they went to sleep at night. He traces how guards and doctors kept ledgers of events and behaviours to enable the sharing of information. In this way, he argues, prisoners were subject to normalising judgments concerning right and wrong behaviour, which were orchestrated by guards and newly emerging professionals, including doctors, psychiatrists, sociologists and criminologists, all of whom observed, recorded, collated and categorised knowledge of inmate behaviour. Deviation from the norm was thus identified, recorded and corrected, often using the fruits of early scientific research and experiment. For example, the results of human anatomy dissections, particularly from the nineteenth century onwards, were used by some early penal experts, notably Lombroso and the Italian school of criminology, to identify how some criminals were a distinctive human type, who were abnormal biologically to law-abiding citizens and hence needed to be subject to particular disciplinary regimes.

For Foucault, the surveillance gaze was asymmetrical as those subject to it were unable to challenge or resist it. The sense one gets from reading Foucault is that acts of resistance may well be possible in some limited sense, but for obvious reasons ultimately power lies with those doing the surveying. The fact that surveillance utilised the fruits of emerging modern sciences, such as medicine and the social sciences, to justify its practices, alongside the fact that it was by and large enacted on the poor and socially excluded, contributed significantly to this state of affairs. In short, the form of surveillance in place within the confines of the prison walls operated through the panoptic principle – where the few (guards, doctors and a growing range of penal experts) could exercise control over the many (the prisoners). This ensures a new kind of legitimate authority which sought a more intensive, constant, efficient and somewhat automatic functioning of power and control over those individuals subject to its gaze. As we shall discuss in more detail shortly, it is this form of power that Foucault argues has stayed with us. Indeed, he holds it gradually dispersed itself throughout society from the mid-nineteenth century onwards. But for the moment, it is important to note that the development of the modern prison from the mid-eighteenth century onwards acted as a laboratory within which a range of experts and penal reformers experimented to identify how best to monitor, regiment, train and correct individual behaviour, and furthermore, they drew on the emerging fruits of modern science to help them achieve their goals.

Yet it must be recognised that these events are not to be viewed negatively: Foucault notes that many of the new regimes introduced were born out of good intentions and a growing belief that individuals were not fundamentally flawed and indeed could be rehabilitated. Additionally, as Foucault is particularly concerned to point out, what is important about this shift towards the exercise of a more disciplinary form of power is that it is deliberately designed to encourage prison inmates to conduct self-surveillance in regulating and disciplining their own behaviour.

For Foucault, panoptic power and the forms of expertise and unverifiable surveillance this gave rise to do not end at the prison walls. Rather, this distinctly modern form of corrective disciplinary surveillance gradually came to operate outside the prison as a new instrument of social control which would *'insert the power to punish more deeply in the social body.... [as it pursued]...the utopia of the perfectly governed society'* (Foucault 1979:198). Indeed, he argues that by the end of the nineteenth century we can see it operating across society's social institutions, including schools, military institutions, hospitals, asylums, governing bodies and even worker factories. Hence, for Foucault, surveillance is a tool for ensuring obedience which proliferated throughout a range of social institutions such as schools, hospitals, workplaces, army barracks and asylums. In a very real sense then, the disciplinary society is the surveillance society. For in such a society, according to Foucault, the judges of normality are everywhere. We live in a society of the teacher-judge, the doctor-judge, and the social-worker-judge. In this sense, the thrust of modern surveillance is conjoined with the generation of knowledge used to normalise individual bodies, gestures, behaviours, aptitudes and attainments. Accordingly, what once constituted a drive to control illegalities and crime became applicable to a whole range of behavioural contexts in which 'the norm' reigns supreme and surveillance offers the possibility of countering many different forms of deviance, wherever it occurs.

Panoptic surveillance for Foucault is about whether an individual is behaving as they should in accordance with a social rule. For him, modern surveillance sites – such as the school, the workplace and the town centre shopping complex – aid the maintenance of social well-being and order. Hence, they encourage a docile citizenry who are self-inspecting, self-judging and self-correcting in relation to predominant social expectations and norms. In Chap. 5, we will explore in detail how these expectations and norms are bound up with the neo-liberal form of the enterprise self, that is, the responsible subject-citizen who adheres to certain expectations of who they are and should be, as based around free market processes and ideals relating to the production and consumption of services and goods, in order to explore how this has impacted on the ways in which governing frameworks operate. Hence, it is important at this point to turn to consider the dispersal of discipline in more detail.

The Dispersal of Discipline

For Foucault, surveillance is tied to development of a distinct method of social control – disciplinary power – by which the organisational and social control apparatus of the modern nation-state in western societies has gradually emerged over the last

200 years into its present form. In essence, the risk-saturated social conditions associated with the surveillance society means it is by definition also a normalising society in which the norm of discipline and the norm of regulation intersect. Contemporary reforms in governing bodies being firmly linked to ongoing developments surrounding the need to engender self-surveillance and self-control on behalf of subject-citizens as western neo-liberal democratic states continuously seek to minimise risk and maximise profit. In exploring this point further, it is useful to examine the work of Cohen (1985).

Cohen used Foucault's insights to explore how surveillance generates new practices of social control as communication and information technologies develop throughout the twentieth century and in doing so shape the management of deviance via the introduction of various social engineering initiatives, including community-based social interventions, indeterminate sentencing for offenders and the mentally ill, neighbourhood watch schemes and private security firms, as well as more recently, the growth of CCTV cameras. For Cohen, these initiatives reinforced how the dispersal of discipline has adapted to changing social circumstance ever since it emerged from the prison gates. Indeed, Cohen notes that there has been a move towards informal, private, communal controls outside of the mechanisms of the state: in other words just as disciplinary control moved from the prison to the hospital, to the military yard, to the school and to the factory, it has over time moved from being located firmly within what can be defined quite broadly as the governing apparatus and into everyday areas of social life. This process is a somewhat logical outcome of the fact that a key feature of discipline is that it requires all individuals to engage in self-regulation and inspection, not just those who deviate from the norm. Cohen notes that a key outcome of this expansionist process is that it widens the net of the formal system of control by bringing about an increase in the total number of deviants getting into the system in the first place. Related to this, new types of deviants are created as a thinning of the mesh of the net of social control occurs as it expands in this manner, with the result that we see an increase in the overall level of intervention, including in more traditional forms of institutionalisation such as prisons and psychiatric detention, but also in community sentencing and treatment programmes as well. Here, both old and new deviants become subject to surveillance and control in new ways, including electronic tagging, forced treatment orders and drug and alcohol abuse programmes.

According to Cohen, the dispersal of discipline as a method of social surveillance and control throughout contemporary society during the twentieth century brings with it a blurring of the previous boundaries between the public and the private, as well as formal and informal forms of control. This results in more people getting involved in the control problem of risk identification and management. Accordingly, modern surveillance technology heralds the beginning of a more insatiable processing of deviant groups which is undertaken by new experts in new spatial settings, that is, the computer software engineer and the CCTV operator working with facial recognition software in a city-centre control room. Such processes are bound up with a general the intensification in profiling to identify and manage risk. As Bogard puts it in relation to policing, *'if your skin colour, sex, age, household area, matches the computer profile each officer carries while on duty, you're a target, whether you have actually done anything wrong or not'* (Bogard 2007: 97).

Cohen highlights that it is important to remember that although the features of the process may have changed over time the most fundamental fact about what is going on is that it is much the same as what went on historically. For, like Foucault before him, Cohen traces the bedrock of surveillance to an insatiable governmental need to control a population via classifying and ordering it. Citizens must be identifiable and knowable, so they can be handled appropriately with state intervention into their lives occurring as needed, both directly and indirectly by agencies of social order ranging from the school and the prison to the workplace and the hospital, in order to maintain discipline and social control. This process, it is argued, may well have existed for as long as human civilisation, but it was consolidated in the nineteenth century with the emergence of rational, scientific knowledge, which dramatically expanded our ability to analyse and predict human behaviour and change the natural and social world to our liking. Additionally, it is important to remember that part of the attraction of the promise of scientific knowledge is that it is heralded as value-neutral and objective and therefore is trusted by the mass of people, even if the people who use it are not. It is this need to classify in order to analyse, predict and control that has continued to the present and is woven into the minutia of social life, bringing with it new forms of expertise without which it is assumed we can no longer function as parents, travellers, consumers, workers or sexual beings. Hence, as Giddens (1991) notes, we become increasing reliant on a day-to-day level on a mixture of specialised expert knowledge and self-help manuals, as modernity intensifies and progresses into the risk-saturated social conditions associated with high modernity. A point we will return to shortly when we discuss the consequences of this state of affairs for contemporary reforms in medical governance.

Resisting the Surveillance Assemblage

For Cohen then, as a cornerstone of modern social control, surveillance now operates right outside of the formal punitive system of social control, that is, the criminal justice system. It has become more and more dispersed throughout society to such a degree that it is now present within consumer culture, social welfare and communities, as well as everyday family life. For Cohen, inclusionary social control – the use of supervision of offenders in the community – as well as exclusionary social control, such as the use of prisons, will over time merge to reinforce each other and expand even further into every aspect of society. Also, he thinks we are increasingly going to see pre-emptive forms of surveillance to spot and halt deviance at an early stage. There is certainly evidence to suggest that there is a trend for criminal justice agencies as well as health and social care professionals to focus on profiling risky citizens – that is, those who are a danger to themselves or others – in order to prevent deviance in the form of physical violence, domestic violence, sexual abuse and terrorism. Consequently, the value of the work of Foucault and Cohen arguably lies in the fact that they bring to the foreground the idea of panoptic control and its dispersion throughout society. This, in turn, introduces us to the important idea of an unseen observer – perhaps the core icon of modern Big Brother surveillance imagery – and how they pursue relentlessly, via the innovative use of constantly improving surveillance technologies and the classification of bodies, thoughts, gestures and actions, all with the goal of maintaining social order and managing risk.

Although the value of such insights must be acknowledged, it is also clearly useful to reflect on how human resistance happens under such panoptic forms of control. Foucault famously downplayed human agency. He felt that in the face of the combined forces of language, other people and the governing machinery of the state, individuals possessed very little room to manoeuvre and express resistance. So in some respects does Cohen. There is undoubtedly some obvious truth in both Foucault's and Cohen's tendency to downplay resistance in favour of focusing on compliance as by and large the vast majority of everyday social interaction does serve to maintain social order and the agencies of social control are always at hand to deal with individuals who do not conform. However, there is equal truth in the assertion that a key defining feature of human beings is the capacity to resist and purse social change, while the growth of modern information and communication technologies, particularly mobile phone technology which has undergone a revolution over the last decade, offers us an avenue through which to explore the possibility of resistance. Lyon (2001), for example, discusses resistance in relation to the growth of surveillance technology developments such as biometrics, cyber surveillance and identity cards. Here, he notes how although state surveillance practices have expanded and intensified over recent decades, nevertheless the internet and mobile communication technologies allow citizens to turn the tables and survey powerful elites within society. This is an important point. Because the impression you can get from authors such as Foucault and Cohen is that this is all a one-way street, with people being subject to panoptic power in a one directional form.

The work of Haggerty and Ericson (2006) is extremely instructive here. They argue that in the late twentieth and early twenty-first century surveillance has proliferated and generated greater social visibility. That is, one of the key features of the surveillance society is that citizens can no longer hide and go off the grid: They are tagged and tracked via various means of social sorting as they journey from birth into everyday adult life and onto old age and death (Lyon 2001). Yet, in tune with the risk society, Haggerty and Ericson note how contemporary surveillance transforms social hierarchies, rendering them less rigid as people from all social backgrounds and groups – elite and not so elite – are now under surveillance, not least of all because we all run the danger of becoming risky subjects. Furthermore, they say this is because of the rise of the surveillance assemblage, which they say encompasses the advances and extension in information and data gathering we have witnessed so far. Furthermore, they say that this assemblage is what they term rhizomatic, that is, it is the result of unforeseeable offshoots, interconnections and dispersed flows of data globally across borders and institutions. One way of thinking about this idea is the viral adverts one now finds on the Internet for a range of products and services, including movies, holidays and personal grooming products, which seem to emerge of their own accord.

In terms of looking at individuals and their information, Haggerty and Ericson talk about how nowadays it is difficult for people to maintain their anonymity under the surveillance gaze due to the rise of communication and mobile technology. So they talk about the disappearance of disappearance, and equality through technology, with different social groups of individuals being subject to the surveillance assemblage at different times and places for different ends. For example, police can and do survey and record protestors and their conduct and likewise protestors survey and record the police – both these acts of surveillance can and have led to members of the public and the police becoming subject to legal processes.

The idea of the surveillance assemblage then runs against the views of surveillance as a necessary top-down tool of the powerful, which is evident in the institutionally fixed conceptions of panoptic surveillance. This new surveillance, gathered around the use of techniques such as the Internet and mobile phones, allows for the scrutiny of the powerful by both institutions and the general population. They are a number of related consequences of this view of surveillance. First, privacy may be ending for more and more social groups, but that privacy is being traded by subjects for benefits, services and rewards offered by surveillance bodies – for example, when we allow supermarkets to gather information on us via supermarket loyalty cards in return for better deals and targeted vouchers and such like. This also shows how surveillance is increasingly being rationalised around seduction. In other words, the acceptance of surveillance as part of our lives is greater not because of any directly oppressive or coercive reason, as associated with more traditional panoptic power, and institutionalised in the form of the prison, but rather because we have become seduced to conform to the pleasures of consuming goods offered by the corporate bodies which survey our everyday shopping habits.

Secondly, the assemblage allows for greater expandable mutability as surveillance regimes intended for one purpose find themselves used for another. This is evident in the use by the police of non-police databases for fighting and preventing crime and terrorism; such as financial, educational, media or insurance organisational surveillance records pertaining to individual behaviours and habits when they are suspected of being terrorists. Thirdly, the assemblage's supposed rhizomatic nature is arguably underpinning an as yet unfinished democratisation of surveillance. This is because the panoptican, where the few see the many, is being supplanted gradually over time by, or at least joined up with, another equally pervasive surveillance medium – the synopticon. In a world in which surveillance now enables the scrutiny of the demeanour, idiosyncrasies and foibles of powerful individuals, it is no longer merely the case that the few see the many, as in panoptic power, but rather the many come to see the few, which is synoptic power.

There can be no doubt that Haggerty and Ericson's view of surveillance has taken into account current technological advances and its increasing multimedia character, as well as its ability to enable groups to engage in resistance. The growth of mass media and digital and Internet forms have transformed the surveillance landscape and created what some call 'the viewer' society. Here, collective phenomena such as reality TV shows highlight how people have come to accept surveillance and even revel in the spotlight, while perhaps, most importantly, the many, in other words the watching public, are encouraged to watch and judge the few – celebrities, politicians, VIPs, the notorious and the criminals.

The key question we need to ask ourselves is, of course, if this apparent turning of the tables is indeed happening and the growth of surveillance technology across society not only enables the greater surveillance of the population but also the rhizometric levelling of traditional social hierarchies. Whereas Foucault claimed that people were self-governing when they were part of the panoptic society, could it perhaps be said that in today's society by watching the few we establish shared norms and values, lifestyles and our understanding of the world around us and our place in it? Could synoptic surveillance actually simply be an extension or transformation of panoptic forms of disciplinary power? Just what does the future hold regarding how surveillance can be used to maintain social order within the risk-saturated social conditions of high modernity? It is these broader questions which are shaping the thinking of social theorists as they contemplate how the transformations currently underway in today's globalised surveillance society will unfold and impact on our lives. Hence, it is these questions which the sociological study of professional governance must arguably bear in mind when they examine the contemporary developments in the regulation of expertise. It is to this point the chapter will now turn.

Medical Regulation in the Age of Risk

I would argue that the preceding discussion reinforces why the study of medical regulation needs to take into account the broader context when tracing recent developments in the regulation of the medical profession. As both producer and user of the fruits of modern scientific research, medicine has played a key role in the gradual dispersal of discipline throughout modern society. Medicine has formed a key part of the development of the disciplinary apparatus of the state as it has increasingly sought to manage the risks associated with urban planning and environmental management, public health, social welfare and the problem of criminality. The profiling, care and treatment of a range of risk-laden dangerous sub-populations, including the criminal, the mentally ill, as well as the serially violent and sexually abusive, are all subject in one way or another to modern advances in the technology of health-care delivery, medical diagnostic and surgical procedures and pharmacological regimes. All of which have been bequeathed to us by the success of modern biomedical science in conceptualising, surveying, examining and treating the human body as it suffers from a range of illnesses and diseases on its sometimes all too short journey from the maternity ward to the mortuary table.

Chapters 2 and 3 explore in detail the growth of biomedical practice within Europe from the eighteenth century onwards and how this led to the gradual establishment of hospitals and asylums to enable the mass surveillance and treatment of the sick, dying and mentally ill. As part of this discussion, it is noted that as the nineteenth century progressed and biomedicine established itself as the legitimate heir to the old humoral medical tradition it was became necessary, at least for organisational purposes, to legally establish an approved register of licensed medical practitioners. The resulting model of medical governance, known as professional self-regulation, was patterned on the elitist club governance approach to state organisation which dominated the political classes of the time. In this manner, Chaps. 2 and 3 provide a basis from which Chaps. 4 and 5 can critically examine sociological approaches to the study of professional regulation, both in general and in relation to medical self-regulation in particular. One of the key conclusions reached as a result of this review is that contemporary reforms in how doctors are governed are bound up with broader shifts in how 'good governance' is conceptualised and enacted within contemporary society.

In summary, a key theme explored in the following chapters of this book is how the emergence of modern medicine in closely tied up with the development of modern society and the form its governance takes. The chapters explore how within western nation-states over the last 30 years a neo-liberal economic imperative has sought to increasingly subject the medical profession to greater surveillance and performance management as it has sought to bring medical governance into line with an emerging broader pattern of government and disciplinary control. This is due in no small part to the re-emergence of liberalism and the growing ascendancy of the concept of the enterprise self throughout all spheres of modern social life (Gordon 1996). For example, Burchell (1996: 28) argues that neo-liberalism's dual advocacy of the self-regulating free individual and the free market has led to *'the generalisation of an "enterprise form" to all forms of conduct'*. Similarly, du Guy (1996) argues that enterprise – with its focus upon energy, drive, initiative, self-reliance and personal responsibility – has assumed a near-hegemonic position in the construction of individual identities and the government of organisational and everyday life. Enterprise, he concludes, has assumed *'an ontological priority'* (du Guy 1996: 181). Consequently, as Burchell (1993: 275) notes, *'one might want to say that the generalization of an "enterprise form" to all forms of conduct – to the conduct of organisations hitherto seen as being non-economic, to the conduct of government, and to the conduct of individuals themselves – constitutes the essential characteristic of this style of government: the promotion of an enterprise culture.'* Indeed, reviewing National Health Service reform during the mid-1990s, Johnson (1994: 149) noted that *'government-initiated change has, in recent reforms, been securely linked with the political commitment to the "sovereign consumer"'*. With health and social welfare commentators noting much the same as the Millennium brook and we moved through its first decade (White 2011).

It is because of these broader shifts that focusing on the issue of medical autonomy can be problematic. Yes, of course, to a degree it is necessary to empirically separate the medical profession from the state if one wishes to examine the impact of health service or regulatory reform on both the practitioner and service use experience of health service delivery alongside the performance management and quality assurance of medical work. But at the same time, it is necessary to recognise that the specialist expertise possessed by the medical profession in a very real sense forms part of the disciplinary state governing regime and as such attempts to reform its governance do not result in a decline in autonomy but rather a further transformation of it. Currently, as is argued in this book from Chap. 4 onwards, a good way of characterising this transformation is by saying that structurally the medical profession is being restratified into more pronounced elite and rank and file segments as the need to adopt the rhetoric of an open and accountable governance takes hold in light of an increasing focus on promoting audit and appraisal to minimise risk, maximise consumerism and promote the enterprise form. A key consequence of this is the need to recognise that medical autonomy and professional regulation are not end states, but rather are ongoing processes, which are tied up with the broader ebb and flow of the interplay between formal and informal forms of social control and panoptic and synoptic surveillance, as the dispersal of discipline continues to enmesh the population within its every widening net.

This leads us into a final introductory point which we will return to in the final chapter after our journey through the world of medical regulation is completed over the next several chapters. One of the key themes running though this book is the idea that as we enter into the complex, surveillance-saturated, social conditions associated with the risk society, then an element of mutual distrust edges into being in the relationship between experts (such as doctors) and the general public, as well as the relationship between the governing and the governed. It may well be a truism that contemporary society relies on experts to survey and risk-manage the population, but it is equally true that modern technology frees us from the need to rely on them as we become ever aware of the fact that nobody, not even that most trusted of public servants the medical practitioner, can offer us certainty and a risk-free existence. We are living in an age where the work of all experts, and all forms of authority, is questionable. In these circumstances, it is arguable that the commentator interested in the study of the regulation of professional forms of expertise must consider the type of citizen and forms of subjectivity promoted and sustained by the governing regimes of the risk society (Peterson and Bunton 1997). For it is arguable that under the guise of advocating personal freedom and minimal forms of government as the 'natural' way of things, liberal mentalities of rule run the risk of promoting a highly limiting view of what it is to be a human being, let alone a good citizen, within today's increasingly complex social world.

Conclusion

This chapter has focused on providing an introduction to the core themes running through this book against the background of a broader concern with the changing social conditions in which we find ourselves as the governing conditions of risk society continue to develop at a pace that is unrivalled in human history. The intensity of this pace is due in no small part to the continuing rapid development of communication and information technologies, ranging from the mobile phone to social networking sites and the globalised electronic news media. These are arguably transforming in unforeseeable ways how individuals view themselves, their intimate relationships with others, as well as seek to reduce the threat of natural and human-made threats and dangers to their personal being. In covering this background material, this chapter has introduced the reader to some of the key themes that run throughout this book. Perhaps, the most important among these is that medical profession is undergoing a process of internal restratification as its regulatory regime changes to accommodate the governing needs of the risk society. The role of biomedicine in producing and managing risk, as well as the growth of an element of mutual distrust in the profession-public relationship, will also be explored in greater detail in subsequent chapters. As I said at the beginning of this chapter, my goal in this introduction has been to provide a starting point from which the reader can return to remind themselves of where they began as they engage with the material discussed in subsequent chapters. Hence, before moving onto the next chapter, they might like to consolidate what they have learnt by completing the following self-study activities.

Self-Study Activity

1. Write a short 1,000-word essay which outlines and critically considers the argument that modern surveillance technology may appear to grant individual's greater personal freedom and the opportunity for democratic political dissent, but in reality subjects them to a more intensive form of self-regulating social control.
2. Produce a 10-min PowerPoint presentation which highlights the important role medicine as a profession plays in managing risks to individuals, local communities and nations.

References

Armstrong, D. (1983). *Political anatomy of the body medical knowledge in Britain in the twentieth century*. Cambridge: Cambridge University Press.

Beck, U. (1992). *Risk society: Towards a new modernity*. London: Sage Publications.

Beck, U., & Beck-Gernsheim, E. (2002). *Individualization: Institutionalized individualism and its social and political consequences*. London: Sage Publications.

Bogard, W. (2007). Surveillance, its simulation and hypercontrol in virtual systems. In S. P. Hier & J. Greenberg (Eds.), *The surveillance studies reader* (pp. 203–233). Milton Keynes: Open University Press.

Burchell, G. (1993). Liberal government and techniques of the Self *Economy and Society, 22*(3), 267–282.

Burchell, G. (1996). Liberal government and techniques of the self. In A. Barry, T. Osborne, & N. Rose (Eds.), *Foucault and political reason* (pp. 133–167). London: University College London Press.

Cohen, S. (1985). *Visions of social control: Crime, punishment and classification*. Cambridge: Policy Press.
Cohen, E. (2010). *State and Crime*. London: Sage Publications
du Guy, P. (1996). *Consumption and identity at work*. London: Sage Publications.
Foucault, M. (1979). *Discipline and punish: The birth of the prison*. London: Allen Lane.
Foucault, M. (1991). The Ethic of Care for the Self as a Practice of Freedom: An Interview *in Bernauer, J and Rasmussen, D (1991, editors)* The Final Foucault Cambridge MIT Press.
Fukuyama, F. (1992). *The end of history and the last man*. London: Penguin.
Giddens, A. (1990). *The consequences of modernity*. Cambridge: Polity.
Giddens, A. (1991). *Modernity and self-identity: Self and society in late modernity*. Cambridge: Polity.
Giddens, A. (1999). Risk and responsibility modern. *Law Review, 62*(1), 1–10.
Gordon, C. (1996). Foucault in Britain. In A. Barry, T. Osborne, & N. Rose (Eds.), *Foucault and political reason* (pp. 167–184). London: University College London Press.
Haggerty, K. D., & Ericson, R. V. (2006). *The new politics of surveillance and visibility*. Toronto: University of Toronto Press.
Johnson, T. J. (1994). Expertise and the state. In M. Gane & T. J. Johnson (Eds.), *Foucault's new domains* (pp. 67–83). London: Routledge.
Lupton, D. (2011). *Medicine as culture* (3rd ed.). London: Sage Publications.
Lyon, D. (1993). *Surveillance studies*. London: Sage Publications.
Lyon, D. (1994). *The electronic eye: The rise of surveillance society: Computers and social control in context*. London: Sage Publications.
Lyon, D. (2001). *Surveillance society: Monitoring everyday life*. London: Sage Publications.
Macpherson, C. B. (1962). *The political theory of possessive individualism*. Oxford: Clarendon.
Mythen, G. (2004). *Ulrich beck: A critical introduction to the risk society*. London: Pluto.
Peters, M. (2001). *Poststructuralist, Marxism and NeoLiberalism: Between theory and practice*. London: Rowman and Littlefield.
Peterson, A., & Bunton, R. (Eds.). (1997). *Foucault health and medicine*. London: Routledge.
Polanyi, M. (1967). *The tacit dimension*. London: Routledge.
White, K. (2001). *The early sociology of health*. London: Routledge.
White, K. (2011). *The early sociology of health*. London: Routledge.

Chapter 2
Biomedicine, Medicalisation and Risk

Abstract This chapter outlines the historical emergence of biomedicine and in doing so traces the development of medicine's dominance of definitions and treatments surrounding illness and disease. It discusses how the Greco-Roman humoral medicine tradition formed an uneasy alliance with Christianity and shaped the social regulation and self-disciplining of the body across Europe up until the seventeenth century. It then traces the birth of biomedicine to the enlightenment during the eighteenth and nineteenth centuries. This emphasised the application of reason and science to the government of society and the identification and management of natural, physical causes for illness and disease. This chapter outlines how the emergence of biomedicine led to the birth of the clinic which transformed the nature of the doctor-patient relationship. The key achievements of biomedicine are subsequently discussed before the emergence of postmodernism is outlined and how the identification and management of risk has come to occupy centre stage within contemporary governing regimes. Here this chapter discusses how the medical profession came to be increasingly viewed during the twentieth century as a key agent for social surveillance and control. Attention is focused on medical expansionism and how medicine has increasingly annexed areas of everyday life previously subject to moral, religious and legal control and in doing so medicalised a seemingly ever-increasing range of human behaviours – including human reproduction, mental illness, antisocial behavioural disorders and criminality – in order to subject them to medical surveillance, intervention and control. This chapter provides a necessary historical and conceptual background to the discussion in Chap. 3 on the institutionalisation of medical dominance and autonomy in the form of the principle of professional self-regulation and in Chaps. 4 and 5 on sociological approaches to the study of professional regulation. End-of-chapter self-study tasks are provided so the reader can engage in further study in relation to chapter contents.

J.M. Chamberlain, *The Sociology of Medical Regulation: An Introduction*,
DOI 10.1007/978-94-007-4896-5_2,

Introduction

As was noted in Chap. 1, the study of medical regulation is informed by two overlapping sociological sub-disciplines: the sociology of the professions and the sociology of health and illness. Indeed, it is necessary to begin our investigation of the governance of medicine by focusing on an analytical concept that is shared by these fields of study and which has fundamentally shaped sociological analysis of medical regulation: biomedicine. As we shall discuss, biomedicine is commonly defined as a form of clinical medicine that is based on the disciplines of anatomy, physiology and biochemistry. Although we are nowadays used to thinking of medicine as a scientifically informed practical discipline, this is in fact a relatively recent development. Biomedical knowledge and expertise gradually emerged during the latter part of the eighteenth century.

It is commonly held by sociologists and medical historians alike to be the foundation stone on which the fledging modern medical profession increasingly organised itself from the early nineteenth century onwards. Not least of all because biomedicine's emergence enabled traditional professional groupings, such as the royal colleges, to secure the medical marketplace in the face of competitors while at the same time altering the nature of the doctor-patient relationship in their favour. For example, as we shall discuss in Chap. 3, biomedicine's emergence in the United Kingdom enabled medical practitioners to successfully lobby government for the institutionalisation of the principle of professional self-regulation through the establishment of the General Medical Council (GMC) in 1858. As we shall see in subsequent chapters, when looking at the establishment of the GMC, it is important to account for the role played in its formation by the 'members only' club-like form of state governance which dominated during the Victorian era. The principle of professional self-regulation has shaped debate surrounding who should regulate doctors (and how) for over 160 years. Hence, the purpose of this chapter in examining the birth of biomedicine is twofold: firstly, to set the scene for subsequent discussion in Chap. 3 on the institutionalisation of medical expertise as a self-regulating professional body, here using the United Kingdom context and the establishment of the GMC as an illustrative case study, and secondly, to provide an intellectual backdrop against which Chap. 4 can discuss the different conceptual frameworks employed by sociologists to theoretically investigate entrenched medical power and contemporary challenges to it in the form of threats to principle of professional self-regulation and individual practitioner's clinical autonomy at the patient bedside. In achieving these goals, this chapter emphasises the historically and socially located nature of modern medical expertise and, by extension, its regulation. But before we can do this, it is necessary to begin with an idea that is as much foundational to the practice of modern liberal democracy as it is to the notion of scientific discovery: progress.

The Enlightenment Promise

Progress. If I were to ask you to think for a moment about the concept, a mixture of words and mental imagery relating to a person, event or thing undergoing a discernible series of step-by-step stages will more than likely spring to mind. So you perhaps may think of your own birth and how over time your body and sense of self has developed and changed as you progressed from being a baby to a small infant, through adolescence to your early teens, and on to the age you now are. Alternatively, you may associate progress with improvement. Perhaps here thinking about how the first computers started out as massive room-filling machines before becoming smaller and smaller, and at the same time faster and faster, until they can now sit comfortably in your lap, or your hand even. We can think of medicine in such terms. Looking back of the history of the development of how medical practitioners treat illness and disease reinforces how ideas concerning what the causes of many human aliments are, alongside the technologies and interventions used to diagnosis and treat them, have progressed over time so we are now able to mass produce stocks of preventive vaccines and treatment medicines, saving millions of lives as we do so. Here then, the idea of progress, medical and otherwise, can be said to be bound up with notions of change, development and, perhaps above all, improvement. In modern times, we often talk about progress as if to imply some sort of improvement has taken place. So we can talk about how progress has been made when we look at the social history of women and voting rights, or when the new iPhone is released, or when a previously underperforming athlete or football team wins a competition. Yet this association of progress with improvement is a relatively new occurrence.

Encyclopaedias of human history like to portray the emergence of the modern world we now live in as a gradually unfolding civilising process involving technological and social progress from the time of our cave-dwelling hunter-gatherer ancestors, to the early city-dwelling worlds of ancient Egypt, Greece and Rome, and subsequently onto the early birth of modern European nation-states as the dark ages of the fall of the Roman Empire moved into the middle ages of Christian piety, before turning to consider the rest of the world and the European exploration of the west, in the form of the discovery of North and South America, and the East, with the (re)discovery of India, Africa and China. Yet, in fact, for much of human history the great mass of people perceived the art of living as an inherently cyclical affair by its very nature, with the perpetual turning of the seasons reminding them of their preordained, unchangeable lot in the world. Just as the seasons turned from spring, to summer, to autumn and winter, so too childhood and youth turn to adulthood, which in turn eventually give way to old age and death. Much of early religious and spiritual thought reflects shared recognition of the deep circular linkages which exist between the turning of the seasons and the unfolding of an individual human life.

Of course, as recorded human history developed from around 500 BC onwards, people could increasingly see that as generations passed by improvements in living conditions were emerging. How could they not as the ancient city-states of Athens or Rome flourished? Yet the limited and precarious nature of the progress achieved was all too apparent, with a lethal mixture of disease, pestilence, famine and war together acting as constant reminders that all things which grow eventually wither and die, including cities and empires. Additionally, religious and political elites were readily on hand to remind the masses of their ordained place in the scheme of things should they forget or, heaven forbid, seek to challenge entrenched social order. For most of human history, people lived lives that more often than not were brutish, violent and short, and the idea of progress as improvement – be it expressed in terms of the improvement of self or the improvement of one's social circumstance – was the preserve of an elite few, and even this was tempered by an acknowledgement that the ability of human beings to change the world around them was severely limited.

By the middle ages, any sense of human progress was further limited and shaped by a European worldview dominated by predetermining Christian notions of original sin, heaven and hell, as well as the divine right of kings and queens to rule with an iron fist. Indeed, it is arguably not until the emergence of the intellectual point in history known as the enlightenment during the seventeenth century that a sense of progress in terms of improvement began to finally emerge in a form we would understand today. For a key promise of the enlightenment was that humankind could change their circumstances for the better, both individually and collectively. This is why the enlightenment is held to signal the beginning of the age of modernity. The engine room driving this age of progress was the rationalistic, experimental methodology of the scientific method. Without it, as we shall see, biomedicine would not have come into being.

The enlightenment age sprang into life during the seventeenth and eighteenth centuries. Yet the seeds for its development had been sown some 2,000 years previously with the emergence of ancient Greek Philosophy, which is often said to have come into being with the prediction of an eclipse by Thales of Miletus in 585 BC. Greek philosophy advocated the development of the questioning mind instead of simply relying on cultural tradition and religious doctrine to guide the way. Hence, it emphasised the application of human reason and practical experiment to uncover the hidden structures and patterns which seem to control our environment. True, Greek thought, suitably mixed with a large dose of Roman stoic pragmatism, had remained influential throughout the intervening centuries, at least in the well-educated upper classes of European polite society, but it was not until the right social conditions emerged that the great transformation could begin. Of particular importance here was the reformation in the sixteenth century which eventually led to the splitting of the Christian church into Protestant and Catholic parts. The gradual turning away from the Holy See of Rome, along with Protestantism's emphasis on the individual and the personal pursuit of God's love through a mixture of a puritan work ethic and freedom of action and worship (albeit within strict social mores regarding 'good conduct'), together opened up a cultural and intellectual space for alternative philosophies to flourish.

Key founding intellectual figures for the enlightenment included Spinoza (1632–1677), Locke (1632–1704), Newton (1643–1727), Franklin (1706–1790), Voltaire

(1694–1778), Rousseau (1712–1778) and Montesquieu (1689–1755). The centre of the intellectual explosion was France, but it quickly found its way to England, Scotland, Ireland, Germany, Holland and into Eastern Europe. The enlightenment created a social environment conducive to scientific experiment, and this, in turn, drove the development of the industrial revolution from the nineteenth century onwards. Enlightenment calls for liberty and freedom of thought became tied up with an emphasis on reason and science as the key tools for pursing both individual and social progress. While democratic ideals and calls for freedom from repressive forms of monarchial rule found expression in the American Declaration of Independence (1776), the United States Bill of Rights (1789), as well as the French Declaration of the Rights of Man and of the Citizen (1793).

A Realist Endeavour

> The primitive does not distinguish between medicine, magic and religion. To him they are one, a set of practices intended to protect him from evil forces.
>
> Sigerist (1951: 127)

Enlightenment modernism heralded the emergence of the belief that progress could be made towards a utopian social order through controlling the physical and social worlds via the application of scientific rationality. It was held that as scientific forms of knowledge developed so would the ability to solve common social problems such as hunger, disease, crime and poverty. As we will shortly discuss, the growth of modern biomedicine was firmly bound up with the enlightenment project. Key to this endeavour was a realist conception of reality. Here, objective reality is held to exist independent of our perception of it. The realist position appears commonsensical. As we go about our everyday lives, we typically assume the world around us has existed since before we were born and indeed will continue to exist after we die. We also notice it possesses regular forms, patterns and processes, with it being the goal of science to uncover the underlying forces behind them. What is more, human beings who live, work and play together tend to possess shared values and beliefs of the nature of the world in which they live that guide how they interact with each other, in part because these shared values and beliefs are internalised by individuals from a young age through the processes of socialisation. Furthermore, these objectively and materially confront us on a day-to-day basis as external social facts in the form of social organisations and institutions which embody communally shared values and ideals that act to channel individual human behaviour in socially acceptable ways.

The increasing use of the scientific method from the eighteenth century onwards, to examine both the material and social worlds through a realist lens, led to the rapid development of a range of new sciences, including biomedicine, public health, sociology, psychology, psychiatry and criminology, to name but a few. Over the last two centuries harnessing the power of science in this manner has enabled human beings to collectively act to change their surroundings, mass produce crops and livestock to sustain growing populations, eradicate certain diseases and contain still others, build extraordinary cities, as well as develop amazing information and communication technologies which make the global truly local. In short, the realist conception of a single objective reality and the methodological use of science to access its underlying processes and rhythms has arguably enabled the enlightenment promise of progress as improvement to be kept. Yet, as we will discuss later in this chapter, the enlightenment endeavour and realist view of reality is not without its critics. But first it is necessary to explore the birth of biomedicine and to do that requires we go back in time to examine the Greco-Roman humoral tradition.

The Greco-Roman Humoral Tradition

> Certain basic physiological concepts and associated therapeutic methods – notably humoral theory and the practice of bloodletting to get rid of bad humors – had a continuous life extending from Greek antiquity into the nineteenth century.
>
> Siraisis (1990: 70–71)

Biomedicine can be characterised as a scientific system of medical practice which from the seventeenth century onwards replaced dominant religious and magical ways of thinking about and dealing with health and illness as part of a much wider shift to rationality associated with the enlightenment and the 'march of progress' this brought with it in the form of technological and social advancement, industrialisation, urbanisation and democratisation. Porter (1997:9) notes that the *'medical history of humanity in Europe (from Greco-Roman antiquity onwards) can be depicted as a series of stages which, broadly speaking, involved the systematic replacement of transcendental explanations, positing instead a natural basis for disease and healing'*. The age of rationality and scientific endeavour created by the enlightenment brought into being new ways to manage the embodied world of human flesh. Yet since the dawn of recorded human civilisation, the management of bodily ills has lay at the forefront of the noble arts of good governance and healing. Within European philosophical and religious thought the human body has been regarded simultaneously as an object of wonder and as problematic, dangerous and threatening. After all, it attracts illnesses and diseases and contains passions and pleasures, which cause not just individual pain and suffering, but spread social ills as well.

Concern with a range of infectious diseases, including cholera, smallpox, the plague and yellow fever, to name but a few, lay behind the early development of treatises on the care and management of the body, both individual and social. Self-control and self-discipline over the body to minimise the social effects of the spread of infectious diseases, alongside other potential bodily ills such as anger, vanity and excessive passion, formed an essential part of early Greek philosophical treatises concerning the care of the self and the governance of the general population (Lindeman 1999). Hence, Greek philosophers warned of the urgent need for the would-be philosopher to develop the ability to distinguish between necessary and unnecessary bodily wants and needs through the regulation of pleasures. It being recommended that *'we should try and live a frugal life in which necessary desires are satisfied, and natural but not necessary desires given some place, while vain desires were outlawed. Such a life would naturally be virtuous'* (Huby 1969: 67).

The development of early medicine was bound up with this emphasis on living a virtuous life. Within the Greco-Roman humoral tradition, the world possessed four seasons, summer, spring, autumn and winter; consisted of four key elements, fire, earth, air and water; and had four qualities, hot, cold, wet and dry. It was said that individuals had four humours: black bile, yellow or red bile, blood and phlegm, as well as four personality types: sanguine, phlegmatic, choleric and melancholic. Having too much phlegm made you overly heavy and slow, too much blood made you more sanguine, too much yellow or red bile made you more quarrelsome, while having too much black bile made you more melancholic (Temkin 1973). Good health rested on the proper balance of a person's four humours with their personality type. Illness and disease came about due to their imbalances. Diet and physical exercise regimes were recommended to maintain balance. Standard therapies to readjust imbalances included the use of steam saunas, inducing vomiting, hot cupping and bloodletting (often using leeches). Careful examination of the evacuation of sweat, urine and faeces enabled a doctor to diagnose and/or prevent imbalances. Humoral medicine heavily focused on prevention. As Lindeman (1999: 10) notes, *'in humoral medicine, prevention...was as important as treatment. The best means of maintaining health was to practice moderation in all things, especially in the use of...(1) air, (2) sleep and waking, (3) food and drink, (4) rest and exercise, (5) excretion and retention and (6) the passions or emotions. A healthy regimen was predicted on observing these rules of nature and avoiding exhaustion, overheating, overeating, excessive consumption of spirits, and immoderate desires. Such ideas were prevalent, and informed not only medical theories but more popular versions of health and illness as well.'*

The Greco-Roman humoral tradition emphasised how a healthy body was maintained via humoral notions of balance and the self-regulation of lifestyle while also stressing the material and natural causes for disease. Indeed, it perceived disease to be an environmental but ultimately an individual humoral phenomena. Being in the sun for too long when one's humours preferred mild weather could cause imbalances to occur so steps to remedy the situation needed to be undertaken, particularly if hot weather could not be avoided. It is important to remember that the Greco-Roman humoral tradition was influenced by early Greek philosophy and its emphasis on reason and experiment rather that supernatural explanations: Its focus was on the natural world and it was argued that this could be explained in terms that did not refer to anything beyond nature itself. This early form of through-going materialism could be found in the writings of two key founding fathers of the Greco-Roman humoral tradition: Hippocrates (460–370 BC) and Galen (AD 129–217). Take the example of epilepsy. This had been referred to from ancient times as the sacred disease, with its associated fits, visions and forms of talk together being interpreted as meaning the bearer had been possessed by a supernatural spirit. Yet the Hippocratic treatise *On the Sacred Disease* (400 BC) rebutted such superstitious beliefs:

> I do not believe that the "Sacred Disease" is any more divine or sacred than any other disease but, on the contrary, has specific characteristics and a definite cause. Nevertheless, because

> it is completely different from other diseases, it has been regarded as a divine visitation by those who, being only human, view it with ignorance and astonishment.
>
> Chadwick and Mann (1950: 1)

Hippocrates is, of course, famous for establishing medicine as an ethical discipline and the Hippocratic Oath, which binds the swearer to do no harm, is a well-established cultural and ethnical behavioural norm that nowadays is used internationally by medical practitioners to regulate in-group behaviour, promote professionalism and guide clinical judgement. Chapters 3 and 4 examine how medicine has promoted itself as a value-neutral philanthropic discipline providing a disinterested service to the needy with absolute integrity and honesty. In no small part, this is because Hippocrates successor and the father of western-learned medicine, Galen, similarly stressed that a good physician must promote the art of healing by acting ethically and with utmost respect for human life. Yet although the Greco-Roman humoral tradition dominated European understandings of health and medicine, it did not replace supernatural and religious explanations for illness and disease. Indeed, it is generally held that its emphasis on material explanations for ill health and disease was tolerated, or rather more precisely, Christianised. To be sure, as Turner (1995: 20) notes, *'there was considerable conflict between the secular assumptions of Greek Medicine and the spiritual aims of Christian religious practice'*.

Christianity valued humoral medicine as a worldly practical discipline, but the ultimate authority was ecclesiastical. Yet there was considerable congruence between the Greco-Roman humoral tradition and Christianity. Particularly, given humoral medicines focus on the moderation of bodily desires so the individual could take responsibility for their humour and in doing so live a virtuous life. This chimed with the Christian emphasis on hygiene, cleanliness, dirt and, above all, sin. In short, both the Greco-Roman humoral tradition and Christianity operated within a moral discourse which promoted a set of practices for the regulation of the body and the mind (and the Christian concept of the soul) at the level of the individual and the population.

It was certainly the case that a key dynamic factor running through the development of medicine in Europe up until the early emergence of biomedicine in the eighteenth century was the tensions which existed between the more secular Greco-Roman humoral tradition and the spiritual rules of Christianity. Yet perhaps somewhat ironically it was the firm emphasis placed by the enlightenment on material processes and explanations, as obtained through the application of reason and experiment, which gradually led to the decline of the humoral worldview. As Turner (1995: 29) notes, *'Galen's work 'On the Conduct of Anatomies' became the definitive source for medical understanding of the structure and function of the human body until it was successfully challenged in the late sixteenth century'*. The decline of the Galenic worldview started with the Renaissance (Temkin 1973).

The renaissance was a cultural movement which began in late fourteenth century Florence, Italy, that celebrated both the arts and the sciences, albeit under the watchful gaze of the church. Key renaissance artists such as Michelangelo (1475–1564) and da Vinci (1452–1519) familiarised themselves with human anatomy, in doing so reproducing detailed artistic sketches of the human body for the first time in centuries, which highlighted that Galen had actually dissected animals, not humans, when constructing his anatomical principles. Indeed, following da Vinci, the Flemish physician Andreas Vesalius (1514–1564), in his anatomical text *On the Fabric of the Human Body* (1543), provided detailed illustrations which showed that the human breastbone has seven segments, not three as Galen stated (Siraisis 1990). The work of Vesalius reinforced that Galen had experimented on animals instead of human beings and calls began within medical circles for more research to investigate the previously closed space of the inside of the human body in further detail.

It is startling to think that it was not until this time that medical practitioners had thought to question – at least in public – the received wisdom of humoral medicine, particularly given its emphasis on natural processes and their examination using experimentally gained practical knowledge. Yet this in itself reinforces the scale of the social and cultural dominance of the church which existed at this point in human history. Why question the established truth of Galen when ecclesiastical authority was behind it and readily on hard to ostracise (or worse) the doubters of church-endorsed received medicinal wisdom? Yet first the renaissance and then the enlightenment gradually opened up the necessary intellectual and creative spaces in which questioning minds could think freely and challenge established truths.

Taking advantage of this state of affairs to conduct some research of his own, the Englishman William Harvey (1578–1657) showed how the circulatory system worked through demonstrating in a series of public lectures the attachment of veins and arteries and how blood flowed in a circular motion around the body. This directly contradicted the essentially static conception of blood which existed in the humoral tradition. The work of Harvey led to the further questioning of the authority of Galen and there was a flurry of corpse dissection amongst medical practitioners across Europe (Temkin 1973). Another important early contributor was the French surgeon Ambrose Paré (1510–1590) who significantly advanced sixteenth century surgical techniques and battlefield medicine and is rightfully regarded as one of the founding fathers of modern surgery. Paré sought to discover how best to treat and heal wounds by focusing on using dissection to trace the internal damage coursed by weapons, instead of relying on the received Galenic humoral wisdom of the time to treat injury, with his approach winning plaudits from not only other medical practitioners working under the conditions of war, but perhaps more importantly, military and political leaders as well. But it was the work of Philippus Paracelsus (1493–1541) which possibly had the greatest impact in the early development of biomedicine. Paracelsus's dissection work led him to break completely with the Galenic tradition of seeing disease as the result of humoral imbalance. His empirical investigations of diseased corpses laid the foundations for modern medical practice by leading him to conceive of it as an entity – an *Archeus* – which entered the human body (Siraisis 1990). Slowly, gradually, biomedicine was coming into being.

The Birth of Biomedicine

> At the end of the eighteenth century a new type of medicine swept away the old humoral theories of illness that had dominated clinical practice for hundreds of years. The distinctive feature of the new medicine was its claim that illness existed in the form of localized pathological

> lesions inside the body…. The new model of disease – often called biomedicine because it reduced illness to a biological abnormality inside the body – led to enormous resources being invested in the examination of anatomical and physiological processes, both normal and abnormal, to identify the underlying basis of pathology.
>
> Armstrong (1995: 1)

The work of Paracelsus, Paré, Vesalius and Harvey is important as collectively they established the intellectual and technical groundwork from which biomedicine would eventually emerge. Through advocating the use of dissection, they promoted a scientific basis for medical practice which refocused the art of healing away from the rote learning of ancient texts and towards the gaining of direct experience of the origins of disease pathology through conducting anatomical experiments (Siraisis 1990). Although in many ways their collective works represent a fundamental breaking with the Greco-Roman humoral tradition, the fact of the matter is that continuities exist between modern biomedicine and the scientific tradition of empirical experimentation first espoused by ancient Greek Philosophers such as Socrates, Plato and Aristotle, and which, in turn, had inspired Hypocrites and Galen to turn away from religious and populist explanations for illness and disease. The humoral and biomedical traditions share in common a concern with identifying natural, material and physical causes for illness and disease. Yet it is important to remember that the anatomical dissection of human bodies was a topic fraught with cultural and social difficulties and injunctions. Indeed, although it is possible to trace the growing use of dissection amongst medical practitioners from the fifteenth century onwards, in practice this was surrounded by severe moral prohibitions. Today, we are familiar with the use of cadavers in medical training as well as the development of medical understanding of the structure and function of the body and illnesses and diseases which can afflict it.

Yet at the time of the enlightenment both Christian and popular attitudes towards the body actively discouraged its use as an investigative learning tool. An overriding social and cultural concern for the metaphysical soul, whose care involved controlling bodily urges and desires, meant that the body was viewed with a mixture of fear, loathing and contempt, while at the same time being subject to a series of ecclesiastical processes, symbols and rituals to ensure the souls entry into heaven (Turner 2008). It was seen as essential for the deceased to be kept intact and subject to appropriate washing, prayer and burial rituals, under the supervision of an appropriate qualified member of the church. Consequently anatomical dissection undertaken for medical research purposes was viewed as a punitive exercise by much of Christianised Europe as it could put the possibility of resurrection into the afterlife at risk (Turner 1995). Dissection was associated with punishment and seen by many as a fate worse than death. What is more, it was legally held to be an extension of punishment for criminal behaviour, with the result that more often than not individuals subject to public execution, and who had the financial means to do so, would try to bribe their executioner to ensure they were buried appropriately after they were executed and not sold for anatomical dissection (Richardson 1988). As Turner (1995: 29) notes, *'dissection was an essential part of the whole process of legal punishment and juridical execution'*. For example, in the United Kingdom, the Anatomy Act of 1832 allowed practitioners to obtain corpses from workhouses and hospitals that were unclaimed by relatives, alongside bodies bequeathed to them by members of the public of their own free will.

Before the 1832 Anatomy Act, the only legal supply of corpses for anatomical purposes was those condemned to death by the courts. Yet such corpses were not always appropriate for dissection due to damage suffered during execution. Furthermore, their supply was also somewhat naturally limited. In no small part this state of affairs was why anatomical dissection was increasingly associated by the general public with grave robbing and even murder as it became a more and more popular learning tool amongst medical practitioners during the seventeenth and eighteenth centuries. An infamous example of this was the case of William Burke and William Hare, who between 1827 and 1828 committed serial murders to supply Dr Robert Knox of Edinburgh Medical School with 17 corpses for dissection. Burke and Hare's case was covered extensively in the news media of the time and led to a sudden growth in burial tombs specifically designed to be stop grave robbers. Indeed, it was cases such as Burke and Hare which eventually led during the nineteenth century to growing state intervention across Europe to enable the legitimate acquisition of corpses by medical practitioners for dissection, instead of restricting their supply base solely to convicted criminals (Turner 1995). Nevertheless, a collective sense of unease surrounding the cutting open and dissection of the body persisted well into the twentieth century, and indeed, it is arguably still with us today, albeit in a less directly Christianised form.

The dissection of corpses was then a heavily taboo subject during the seventeenth and eighteenth centuries, with the result that the development of biomedicine during this time was a gradual and often secretive affair, with the fruits of anatomical dissection often only being shared within a small circle of practitioners. Public dissections for instructional purposes were extremely rare events and when they did happen they were more often than not accompanied by a range of presentational symbols and rituals, often including a bible reading, a banquet and formal styles of dress, which together served to reinforce the potentially spiritually dangerous nature of the event for observer-participants (Lupton 2011). Such historical circumstances reinforce to the medical historian that biomedicine's emergence was in no small part heavily dependent on broader sociopolitical changes which were occurring during the eighteenth and nineteenth centuries, particularly the gradual assent of liberal democratic, as opposed to monarchical, forms of governance within western nation-states. For this led to a growing separation of religious authority from political authority, with the church over time gradually playing less and less of a direct role in the day-to-day government of society, and political elites increasingly turning to science and an expanding number of associated specialised experts to help them govern; including doctors, sociologists, criminologists and public health specialists to name but a few.

It is important to remember that the emergence of biomedicine (and indeed the modern medical profession) is very much bound up with a growing broader sociopolitical reliance on science and its methodological emphasis on the empirical investigation of natural and social phenomena for the purpose of ensuring the social good.

Furthermore, the work of key influential scientists, such as Galileo Galilei (1564–1642), Blasé Pascal (1623–1662), Gottfried Leibniz (1646–1716), Isaac Newton (1642–1727), Christopher Wren (1632–1723), John Hadley (1682–1744), Henry Cavendish (1731–1810) and Joseph Priestly (1733–1804), all promoted a mechanical, machine-like, view of nature and the universe. Viewing the human body in similar machine-like terms enabled it to be conceptualised as understandable, classifiable and, above all, repairable. The foundation stone on which the machine metaphor was applied to the examination of the human body was pathological anatomy. Xavier Bichat (1772–1802) examined the tissues of organs to ascertain the localised nature of disease. Giovanni Battista Morgagni (1682–1771) used an early microscope and identified the clinical features of pneumonia. Mathew Baillie (1761–1823) accurately described cirrhosis of the liver.

Morgagni, Baillie and Bichat together signify the beginning of medicine's focus on abnormality as much as normality and the concurrent use of morbid anatomy as a methodology to further medical knowledge and practice. Indeed, Bichat is quoted by Carter (1991: 543) as saying, *'open up a few corpses (and) you will dissipate at once the darkness that observation alone could not dissipate'*. Biomedicine had finally arrived. As a result of the increasing use of dissection as a learning tool during the nineteenth century, there was a growing acceptance amongst medical practitioners, as well as society at large, that diseases were caused by specific entities entering the tissues within the body. The first individual to observe bacteria and other microorganisms, using a single-lens microscope of his own making, was Antonie van Leeuwenhoek (1632–1723) in 1676.

The subsequent refinement of the microscope over the next century and a half dramatically increased the identification of bacteria and other disease-causing entities, such as protozoa: The observational evidence was available for anybody who cared to look for themselves. Biomedicine is reductionist and materialistic: It seeks to explain the phenomena of health and ill health in terms of cellular and molecular processes and events. Pathological anatomy and the microscopic examination of body parts enabled medical practitioners to locate disease within the organs and systems of the body. By the middle of the nineteenth century, the intellectual break with the Greco-Roman humoral tradition was virtually complete. Illness and disease were now perceived as objective entities, with external symptoms being increasingly linked to internal processes within bodily organs and tissues, in contrast to the Galenic theory of bodily disturbances and the need to maintain humoral balance.

It is undoubtedly the case that the emergence of biomedicine enabled medicine to do its bit in fulfilling the enlightenment promise of promoting progress as development and improvement for the social good. From the mid-nineteenth century onwards, the biomedical approach towards the body had established the laboratory as the site where the dissection of corpses and the application of new medical technological instrumentation, such as the agar plate, the X-ray and the stethoscope, meant medical research scientists were able to establish the causes of infectious diseases that had haunted civilisation for as long as recorded human history. By the end of the nineteenth century, cholera, tuberculosis, typhoid and diphtheria had been identified, examined and catalogued. Public health specialists used the fruits of biomedical research to ensure deaths rates from such diseases plummeted.

The discovery of penicillin and streptomycin in the first part of the twentieth century led many to believe that influenza and tuberculosis could be eliminated completely. Medicine was increasingly viewed with a mixture of awe and respect by the public.

Given the advancements that have been made by medicine, it is perhaps unsurprising that medical historians and medical sociologists tend to hold that its technological and diagnostic advancements and successes throughout the last century have led to a biomedical discourse dominating contemporary debate surrounding public health, as well as the organisation and delivery of health care (Lupton 2011). But it is important to recognise it was the successful mobilisation of biomedicine as an information resource and investigative tool, collectively by the medical profession for its own ends, which ultimately enabled entrenched medical power and control over healthcare delivery, to develop. We will examine this point in greater detail in subsequent chapters. But as a prelude to that discussion, it is important to consider the birth of the clinic and the emergence of the medical gaze.

From the Laboratory to the Clinic and the Medical Gaze

Biomedicine is a distinctive form of clinical practice which takes the merger of the laboratory and the clinic during the nineteenth century as the engine room for medical innovation, the establishment of institutionalised medical power, as well as the transformation of the doctor-patient relationship. Up until this point in time the patient interpretation of their illnesses was seen to be an important part of the diagnostic and treatment process. Under the Greco-Roman humoral tradition the patient acted as a patron and largely determined the dynamics of the medical encounter as well as the course of treatment (Jewson 1974). The emergence of biomedicine shifted the diagnostic and treatment emphasis away from a patient's personal experience of illness and disease and towards the objectively obtained facts pertaining to a clinical presentation. The person-centred Galenic cosmology of illness shifted to the object-centred cosmology of biomedicine. Jewson (1976: 235) characterises the nineteenth century emergence of biomedicine as the point in the history of medicine where sick individuals became patients who were *'designated a passive and uncritical role in the consultative relationship, his [sic] main function being to endure and wait'*.

Yet the shift to the biomedical model as the source of medical knowledge did not in itself possess the totalising power necessary to irrevocably change the doctor-patient relationship. Rather, it was the concurrent emergence of the state-endorsed hospital clinic as a site for the rational treatment of illness and disease, both physical and mental, which radically altered the relationship between the medical profession and the public. Arguably, it was with the birth of the hospital clinic and the mental asylum that medicine as a profession became a state-sponsored mechanism for social surveillance and control. With the birth of the clinic, medicine came to possess legitimate state-endorsed power over the care and treatment of individual body in addition to significant political influence over the formation of health and social policy due to its apparent objectivity and seemingly scientific nature (Foucualt 1989).

The enlightenment period which overtime gradually gave rise to modern democratic European nation-states was a time of rapid social change and political unrest. Not least of all because it brought with it industrialisation, urbanisation and the early development of globalised free market economics. As a result, there was an increasing concern, on behalf of governing elites, with securing social order under the emergence of the democratic ideal of government by the consent of the people, rather than more traditional notions of absolutist church-endorsed monarchical power. In the face of the growth of democratic idealism, social and political elites were well aware of the necessity of securing the social welfare and public health of the population. In broad terms, it is possible to conceptualise the eighteenth and nineteenth centuries as the point European history were there was a shift from religious and punitive to secular and rehabilitative forms of population-wide surveillance, correction and control (Turner 2008).

The enlightenment signalled the birth of the age of reason, and drawing heavily on the emerging natural and social sciences of the time, nation-states across Europe sought to establish the rational and bureaucratic management of the body through sponsoring the development of secular sites for population surveillance and management. Hence, for example, during this time, there was a shift from the use of public execution to punish criminals, towards the modernisation of the criminal justice system to incorporate rehabilitative ideals and the increasing use of prisons as corrective sites for the (re-)training of the deviant bodies. While as Turner (1995: 34) notes, we also see the *'rise of the hospital as a teaching institution rather than as a general dumping ground for the poor and destitute'*.

The establishment of the hospital as research and teaching site allowed for the surveillance of a large number of patients, the observation of generalisable disease patterns to help refine medical treatment techniques, as well as greater population control via the introduction of public health and urban sanitation measures. Consequently, the individual body was firmly established as a site for social surveillance and inspection as well as the advancement of rational, scientific, medical knowledge. As the hospital increasingly became a location for medical research and training, the body was sampled, measured and generally coerced into revealing its secrets by a growing number of specialist medical disciplines, departments and laboratories. As Armstrong (1983: 2) notes, *'[the] medical gaze, in which is encompassed all the techniques, languages and assumptions of modern medicine...[established]...by its authority and penetration an observable and analyzable space in which...[was]...crystallized that apparently solid figure – which has now become familiar – the discrete human body'*.

The birth of the modern hospital is generally recognised to have occurred in 1790s France before rapidly spreading across the rest of Europe during the early part of the nineteenth century (Foucault 1989). Prior to this time, hospitals were relatively small-scale charitable enterprises, mainly focused around containing the poor and socially dispossessed to secure the salvation of their soul. But the clinic came to be seen as an invaluable governing tool as nation-states accepted the utility of biomedicine for securing public health. A point which can perhaps be best illustrated by the fact that it is possible to trace a rapid rise in clinic patient numbers from some 200 or so per annum during the eighteenth century to over 22,000 per annum (and rising) by the early nineteenth century (Turner 1995). Key to the success of this new governing role was the application of the medical gaze as a diagnostic and teaching tool.

Foucault (1989) argues that within the clinic emphasis was placed on gaining direct personal experience of presenting clinical phenomenon. External, observable signs and symptoms were increasingly matched to the findings of pathological science. Meanwhile, emerging medical technologies, such as the stethoscope, alongside emerging patient examination techniques such as palpation, auscultation and percussion, together reinforced the legitimacy of this new observational approach to medical practice. This laid the foundation stones for the development of modern medicine's 'craft expertise', with a doctor's own direct scrutiny of a patient becoming paramount. Indeed, Foucault (1989: xvii) notes that, *'clinical experience ...was soon taken as a simple, unconceptualized confrontation of a gaze and a face, or a glance and a silent body, a sort of contact prior to all discourse, free of the burdens of language, by which two living individuals are trapped in a common, but non-reciprocal situation'*.

In summary, although the clinic was founded on the quantifiable (and socially and politically useful) nature of scientific medical knowledge and expertise, it was the emergence of the medical gaze which led to the patients' narrative of their personal experience of illness and disease becoming wholly secondary to the doctor's esoteric clinical-anatomical experimental expertise. Chapter 3 discusses how the medical profession utilised this mix of biomedical knowledge and tacit clinical expertise as a resource to entrenched medical power; no more was the individual practitioner to be subject to the whims of their patients. As Jewson (1976: 235) states, *'henceforth the medical investigator was accorded respect on the basis of the authority inherent in his occupational role rather than on the basis of his individually proven worth. The public guarantee of the safety and efficacy of theories and therapies no longer rested upon the patients' approval of their contents'*.

Furthermore, the emphasis on the primacy of tacit clinical knowledge and expertise gained through obtaining direct clinical experience under apprenticeship has been a regular feature of sociological accounts of medical training and practice for the last several decades. Becker et al. (1961: 225), in their groundbreaking account of medical school training, discussed how personal expertise gained from actual clinical experience is often contrasted by clinical teachers to available scientific knowledge, '[*so even] though it substitutes for scientifically verified knowledge, it can be used to legitimate a choice of procedures for a patients treatment and can even be used to rule out the use of some procedures that have been scientifically established*'. Similarly, Atkinson (1981: 19) in his ethnographic study of bedside teaching and learning in Edinburgh comments how medical trainees experience a *'recurrent reinforcement of the primacy of clinical knowledge over "theory"'*. Sinclair (1997) in his more recent study of medical training in London similarly highlights that during clinical training neophyte students encounter an occupational culture which reinforces the primacy of personal knowledge gained through experience. Here, they are told: *'quite explicitly...that they must learn how to think in a medical way, that preclinical teaching has stopped them being able to think and so on'* Sinclair (1997: 223). Chamberlain (2009) discusses how medical practitioners lay claim to tacit clinical knowledge when asked to account for their teaching and clinical practices by third parties, including educationalists and hospital management. This tension between the quantifiable and more qualitative aspects of modern medical expertise will be explored in subsequent chapters in relation to the institutionalisation of medical power in the form the principle of professional self-regulation. For the moment, it is necessary to focus in the rest of this chapter on the biomedical model and medicalisation via discussion of the risk-saturated social conditions associated with the postmodern turn.

The Biomedical Model and the Postmodern Turn

It is often asserted that the pre-eminence of the modern medical profession lies in its scientific knowledge base, and in turn, this is linked to the historical development of pathological anatomy and the establishment of the hospital clinic as a site for the application of biomedicine (Nettleton 2006). It is because of this that medical sociologists often use the terms 'biomedicine', 'medical model' or even 'biomedical model', as shorthand ways of describing the dominant approach to the identification and treatment of illness and disease within western societies. Here, the doctor is viewed as a mechanic who is treating a defective machine with presenting signs and symptoms being related through the application of medical science and the personalised clinical expertise of an individual practitioner, to underlying physical processes and abnormalities. Hence, Williams (2003: 12) states that the biomedical model has *'four characteristics – (i) disease as a deviation from 'normal' biological functioning; (ii) the doctrine of specific aetiology (specific diseases are caused by specific micro-organisms); (iii) the generic or universal nature of disease, regardless of nature, time and place; (iv) the "scientific rationality" of medicine'.*

It is undoubtedly the case that the biomedical model achieved some remarkable successes. Over the course of the nineteenth and twentieth centuries, life expectancy dramatically increased, understanding of what constituted good human diet improved to help promote healthy lifestyle choices, public health medicine transformed sanitation and living standards, while drug therapies and medical treatments led to the dominant killer – infectious disease – being successfully challenged. Yet infectious diseases were replaced by the rise of degenerative illnesses such as coronary heart disease and cancer. This reinforced the limits of biomedical explanations for degenerative illnesses are complex health issues which arguably possess a social and cultural origin as much as they do a biological one. A critique of the biologically deterministic nature of the biomedical approach emerged during the latter half of the twentieth century.

The biomedical model was not without its critics and challengers. Indeed, the second half of the twentieth century was increasingly characterised by growing awareness of risk, medical uncertainty and the limits of modern medicine, alongside growing public distrust in medicine as an ethical profession. It was argued that modern medicine needed to more fully account for the role of society and culture in the creation of illness and disease, not least of all because McKeown's (1979) epidemiological analysis of morality rates in England and Wales reinforced the importance of the role of environmental factors, such as social inequality and poverty, in helping to shape the social distribution of health, illness and disease. Medical sociologists investigated the effects of public sanitation, water supply, nutrition and housing, on social welfare and public health. As a result, they were keen to highlight the role of industrialisation, capitalist economics and socio-economic inequalities, in determining life expectancy and the individual experience of illness and disease. Such developments led to biopsychosocial models of illness and disease gradually emerging from the 1980s onwards (Nettleton 2006). Here, a range of social and psychological factors, including substance abuse, living circumstances, family relationships and mental health matters, are integrated within a biological model of illness and disease to aid the clinical decision-making process and guide a more multidisciplinary approach to public health medicine as well as individual patient treatment and care (Wade and Halligan 2004).

Of particular importance in driving the critique of biomedicine was the feminist social movement as well as the emergence of postmodernism and social constructivist interpretations of reality. Feminist commentators stressed how women's lives and bodies had become increasingly subject to medical definition and control. It was undoubtedly the case that modern medicine had brought with it huge benefits for women when it came to pertinent matters such as childbirth and human reproduction. After all, that fewer and fewer women died in childbirth, or during botched abortions, as a result of the emergence of modern medical advancements and techniques, was undoubtedly a good thing. The birth control pill and birthing anaesthesia undoubtedly dramatically improved the quality of women's lives and, potentially at least, their sense of personal control over their bodies. Yet the actually female experience of who controls their bodies was arguably sidelined by a male-dominated medical profession which seemed to possess an unrealistic but overriding patriarchal concern with, amongst other things, the 'normal' birth (Oakley 1984).

In short, medicine was held to reinforce the culturally gendered notions of females as 'reproducers' with their bodies being reduced to mere baby-making machines, albeit imperfect machines, who were liable to breakdown, which is why medical intervention and observation was said to be needed. After all, women were emotional and irrational; they were 'the other', particularly when contrasted to male-dominated definitions of normalcy, as based around enlightenment notions of reason and rationality. The feminist critique of biomedicine also extended to its realist underpinnings. It was claimed that utopian enlightenment notions of rational progress were inherently masculine and were therefore exclusive towards women who were classified as 'the other' and fundamentally irrational. Consequently, the realism which underpinned the scientific endeavour (and hence the modern biomedical endeavour) was flawed: Reality was viewed through a masculine lens rather than as it really is. Indeed, there was growing scepticism concerning if reality could be viewed as it really is at all.

Here, in its critique of realism and the enlightenment tradition, feminist thought was synergistic with the emergence of social constructivism in the form of postmodernist thought, as associated with the writings of French scholars such as Lacan, Baudrilland, Foucault, Derrida and Lyotard (Butler 2002). Social constructivism argues that realism claims we can obtain accurate and truthful knowledge of the world because the language we use to explore and explain it reflects the nature of things as they are: language is held to transmit reality and is therefore viewed as a value-neutral information-carrying device. A researcher working within the realist paradigm argues that they can obtain knowledge of the world and how it works, including its underlying causal relationships, so they can make predictions which can form the basis for an intervention to change what is happening in the future. Such knowledge is not only generalisable to other contexts it is also value-free and objective, so it is free from researcher bias.

Although they may admit there is truth to the claim that language is an information-carrying device – how could you be reading this book and understand its contents if language does not transmit information and shared meanings? – the social constructivist position nevertheless holds that language is constitutive: It is a site where meanings are created, manipulated and destroyed. Language is not transparent and value-neutral. It is encoded with the power to create reality. In addition to the French postmodernists, the social constructivist turn within the sciences can in part be traced back to the linguistic philosophers Wittgenstein and Winch, as well as the sociologists Berger and Luckmann, who together laid the foundations for an anti-realist social constructivist view of reality whereby accounts of the world are said to not exist independently of the social actor and the language they use to describe the world around them (Philips and Hardy 2002). Over the last three decades, this has led to an increasing focus within the social sciences in particular on the analysis of discourse and the ways language, and associated culturally grounded assumptions and meanings, act to structure ways of talking about a topic, in order to reveal the underlying sociopolitical functions discourse serves. For example, Foucault (1972: 46) noted in his study of the history of madness that *'psychiatric discourse finds a way of limiting its domain, of defining what it is talking about, of giving it the status of an object – and therefore of making it manifest, nameable, and* describable'.

In short, for Foucault, psychiatry creates madness as much as it serves to treat it. Influenced by the work of Foucault, social constructivists have explored how the media, social institutions and specialist forms of expertise act to discursively construct individuals as a locatable object of scrutiny through the different ways discourse is enacted for different purposes in different situations. Consequently, the analytical concern is to examine the discursive resources people draw on, how these came to be culturally available to use, as well as what kinds of subjective positions and social identities they make available and at the same time close off, deny or otherwise silence. Clearly, modern medicine with its public health- and social welfare-oriented focus on the surveillance, prevention and treatment of illness and disease is implicated in such discursive social control endeavours.

As a result of its focus on the constitutive nature of language, one of the key features of postmodernism is its rejection of what are termed 'grand narratives', that is, theories of the physical and social worlds which allow us to apprehend 'truth' and so explain, control and alter them to our will. Biomedicine, with its focus on the biological causes for illness and disease, is arguably one such grand narrative (Rose 1997). In the broadest terms, we can say that they are three interconnected themes running through postmodernism. First, there is a critique and rejection of modernism and the philosophical principles which underpinned the emergence of the Enlightenment period in the eighteenth and nineteenth centuries and saw the development of modern scientific forms of inquiry. As previously discussed, enlightenment modernism heralded the emergence of the belief that progress could be made towards a utopian social order through controlling the physical and social worlds via the application of scientific rationality. It was held that as scientific forms of knowledge developed so would our ability to solve common social problems such as hunger, disease, crime and poverty. From a postmodernist perspective, such faith in progress, science and reason was, at best, naïve and, at worst, deliberately misleading. Yes, technological and scientific developments have over the last 150 years dramatically improved human life expectancy and the general standard of living for some of the world's population. However, such developments have not only brought with them new hazards and risks, such as global warming, but have also remarkably expanded our ability to engage in war and kill each other. While on a global scale social problems such as poverty and hunger remain largely unresolved. Indeed, if anything, the gap between the rich and the poor, the have's and the have not's, appears to be widening.

Second, over the last 50 years, we have seen the emergence of a radically different form of media-saturated globalisation driven by the information and communication technological revolutions. Today's rapidly changing social conditions are characterised by a mixture of liberalist ideology, consumer-driven lifestyle engineering, as well as the increasing questioning of political authority. This has brought about the somewhat paradoxical result that although we have arguably seen a growth in the number of liberal democratic nation-states in the last 50 years, we have also seen the increasing questioning of individuals, organisations and political parties who claim to protect such values. Thirdly, it is argued that the intensification and speeding up of social life – as symbolised by the truism 'think local, act global' – reinforces that there has been a paradigm shift towards an age of uncertainty and an ever-growing public awareness of risk and its mismanagement. There is growing public scepticism of the ability of experts to manage social problems and eliminate risk. Day after day media images of natural and man-made disasters which reinforce the limits of our ability to survey and control dangerousness and risk. This has led to the radical questioning of all forms of knowledge, including scientific forms of knowledge with their claims to objectivity and value-neutrality, just at the point in human history when we are more dependent on scientific and other forms of expertise, than ever before. Subsequent chapters will discuss the impact of these themes on medical governance. For the moment, it is necessary to complete our discussion by turning to examine risk in relation to medicalisation.

Medicalisation

The emergence of an increasing concern with risk and its management against the background of the social constructivist critique of enlightenment modernism of is arguably *the* key to understanding contemporary developments in the regulation of the medical profession. For risk theorists such as Beck (1992) and Giddens (1990, 1991) as contemporary societies – both in western nation-states and the rest of the world – become more complex and diverse, and the technology and communication revolutions continue at pace, more than ever before individuals are required to make life-changing decisions concerning education, work, politics, self-identity and personal relationships. What is more, they must do so in a world where traditional beliefs about social class, gender, race and ethnicity and family are being rapidly transformed. This state of affairs leads to a concern with risk management entering centre stage within society's institutional governing apparatus as well as an individual's day-to-day decision-making (Mythen 2004). Risk theorists argue that throughout human history societies have always sought to risk-manage threats, hazards and dangers. But these management activities have been concerned with natural risks, such as infectious diseases and famine. In contrast to this in today's technologically advanced media-saturated society individuals are seen to be increasingly viewed as both the producers and minimisers of risk. That is, civilisation has developed to the extent that risks are by and large increasingly seen to be the result of human activity. Even events previously held to be natural disasters, such as floods and famine, are now seen as avoidable consequences of human activities, so they must be risk managed. Hence, society's institutions and expert bodies need to become ever more collectively self-aware of their role in the creation and management of risk (Beck and Beck-Gernsheim 2002). While for the individual uncertainties litter an individual's pathway through life to such an extent that everyday events, such shopping for food, buying a new computer or even asking another person out on a date, seem to be loaded with real and potential risks.

One only needs to switch on the television to see how the general public are provided by the news and entertainment media with a seemingly every expanding range of risk-saturated narratives and images. So as they must seek out and engage with on a seemingly ever-growing number of information resources, provided by a myriad of sources, as they seek to navigate their way through their world. Hence, in the risk society, *'[we] find more and more guidebooks and practical manuals to do with health, diet, appearance, exercise, lovemaking and many other things'* (Giddens 1991: 218). And individuals increasingly find themselves having to make 'risk laden' choices *'amid a profusion of reflexive resources: therapy and self-help manuals of all kinds, television programmes and magazine articles'* (Giddens 1991: 20).

Within the risk society, a sense of growing (perhaps even mutual) distrust characterises the relationship between the public and experts. How can we trust what a group of experts are telling us when their advice is often contradictory? Yet, paradoxically, at the same time, a pervasive and seemingly increasingly necessary reliance on an ever-growing number of experts appears to be a key feature of our experience of everyday life. The consumerist rhetoric of freedom of choice may be promoting greater emphasis on the individual as the informed manager of their own life decisions. But at the same time, the disciplinary reach of experts, such as doctors, has arguably expanded to such a degree that they can survey and control human behaviour at distance via factsheets, best evidence guidelines and self-help manuals. Modern medicine is tied up with the governance of contemporary society via its involvement with a number of important social agencies and institutions, including health and social welfare and the criminal justice system. For many medical sociologists, a key consequence of medicine's control over definitions surrounding health, illness and disease is that this allows its practitioners to continually expand their disciplinary jurisdiction into areas of social life previously controlled by law and religion. Medicalisation is one aspect of this expansionism.

Conrad (1992:209) defines medicalisation as *'a process by which nonmedical problems become defined and treated as medical problems, usually in terms of illnesses or disorders'*. It is certainly the case that there has been a seemingly ever-growing number of 'new' medicalised illnesses and disorders over the last several decades, ranging from stress and yuppie flu, to Gulf War syndrome, compulsive shopping behaviour and repetitive strain syndrome. There has been a *'growing penetration of the clinical gaze into the everyday lives of citizens, including their emotional states, the nature of their interpersonal relationships, the management of stress and other lifestyle choices'* (Lupton 2011: 107). Medicalisation reinforces the role played by medicine as an agency for social control. Indeed, since the 1970s, a number of academic writers have argued that medicalisation is not a somewhat natural consequence of medicine being regarded by politicians and general public alike as a successful applied scientific discipline. Medicine has achieved some not inconsequential successes in terms of improving the human condition, particularly in relation to the treatment of illness and disease, human longevity and the promotion of health life-styles. So one might expect it will try to expand its gaze as it continually strives to improve the welfare and health of members of society's members. Yet it was pointed out by some commentators during the latter half of the twentieth century that medicalisation could equally be taken to be symptomatic of the very processes of industrialisation, rationalisation, bureaucratisation and population surveillance which are bound up with capitalist economic system and the perpetuation of social and health inequalities therein.

In short, medicine serves sociopolitical elites as it provides a natural explanation for illness and diseases which may well at least in part be the result of socio-economic and cultural factors. Hence, for medical sociologists such as Freidson (1970), Illich (1976), Navarro (1980) and Zola (1981), modern medicine acts to mask the physical and social harms which often occur as a direct result of the capitalist political economic system, that is, pollution, stress, poverty, war, occupational hazards and so on. While simultaneously removing autonomy away from people to control their own health through making them ever more dependent on the advice of medical experts when it comes to diagnostic and treatment matters concerning illness and disease. As such, the political economy perspective, as it has come to be known, has sought to explore the consequences of medicalisation for sustaining social inequalities and the institutional power and dominance of social elites (Lupton 2011).

We are then in a somewhat contradictory situation. On one hand, the success of biomedicine means the medical profession plays a powerful and influential role in society. A point both the concept of medicalisation and the political economy perspective readily reinforce. Medicine certainly seems every ready to expand its gaze into new areas of social life and human behaviour. The frequent discovery of underpinning biological explanations for an increasing array of human behaviours and mental processes readily attests this point. Yet, on the other hand, the risk-saturated anti-realist social conditions associated with the emergence of postmodernism reinforce the limitations of modern medicine. The tension between these two positions – medicine as solver of risk and medicine as creator (and masker) of risk – will be explored in subsequent chapters as it is arguably the driving force behind much of the contemporary changes which have occurred in medical regulation. It is important to begin this task by looking more closely at how the organisation of the modern medical profession and its regulation is bound up with the formation of broader sociopolitical governing regimes. Consequently, Chap. 3 will pick up the narrative thread laid down in this chapter relating to the emergence of biomedicine to explore its short-term and long-term effects on the organisation and governance of the medical profession.

Conclusion

This chapter has focused on the emergence of biomedicine and its influence in establishing the hospital clinic as a site in which the medical profession could begin to collectively exercise population-wide surveillance and control, particularly from the mid-nineteenth century onwards, as it sought to fulfil the enlightenment promise of engendering social progress through using a scientifically informed form of medical practice to promote public health and advocate healthy lifestyle choices. This chapter concluded by discussing the social constructivist critique of medicine and outlining the medicalisation thesis. Here, it was noted that medical sociologists over the last three decades have increasingly focused on critiquing medical expansionism and the biological determinism which arguably has lain at the basis of modern biomedical expertise ever since it emerged from the dissection room during the seventeenth and eighteenth centuries. This critique has led to the development of biopsychosocial model of illness and disease where the role played by individual psychology alongside social factors, such as gender, race and poverty, in shaping the patient experience of illness and disease is acknowledged. Some of the key themes highlighted in this chapter will be explored in more detail in subsequent chapters, such as the scientific but at the same time deeply personal and tacit foundations of modern medical expertise as well as the contemporary sociopolitical concern with the surveillance and management of risk (including medical risk). The end-of-chapter self-study activities will help you to consolidate what you have learnt before moving on to consider the development of the principle of professional self-regulation in Chap. 3 and the sociological study of medical regulation in Chap. 4.

Self-Study Activity

1. Construct a portfolio of images which over time tells the story of the rise of modern medicine from Hippocrates and Galen to the birth of the clinic and the rapid advances in medical science made in the nineteenth and twentieth centuries. Provide an accompanying account of no more than 1,000 words which critically discusses how your narrative illustrates the power of 'the medical gaze' and its ability to transform our lives for the better while also opening up new possibilities for the surveillance and control of the population by the state and criminal justice agencies.
2. Write a short 1,000-word essay which outlines and critically considers why biomedicine viewed the female body as inherently abnormal, faulty and in constant need of medical intervention. Whose interests do you think were best served by this state of affairs?
3. Produce a 20-min PowerPoint presentation which defines medicalisation and its relationship to risk and postmodernism. Critically discuss the contribution of the concept of medicalisation to the sociological analysis of the biological determinism which arguably underpins the biomedical model.

References

Armstrong, D. (1983). *Political anatomy of the body medical knowledge in Britain in the twentieth century*. Cambridge: Cambridge University Press.

Armstrong, D. (1995). *Outline of sociology as applied to medicine*. London: Arnold.

Atkinson, P. (1981). *The clinical experience: The construction and reconstruction of medical reality*. Farnborough: Gower.

Beck, U. (1992). *Risk society: Towards a new modernity*. London: Sage Publications.

Beck, U., & Beck-Gernsheim, E. (2002). *Individualization: Institutionalized individualism and its social and political consequences*. London: Sage Publications.

Becker, H. S., Geer, B., Hughes, E. C., & Strauss, A. L. (1961). *Boys in white: Student culture in medical school*. Chicago/London: University of Chicago Press.

Butler, C. (2002). *Postmodernism: A very short introduction*. Oxford: Oxford Paperbacks.

Carter, K. C. (1991). The development of Pasteur's concept of disease: causation and the mergence of specific causes in nineteenth century medicine Bulletin of the History of Medicine, 2, 528-548.

Chadwick, J., & Mann, W. N. (1950). *The medical works of hippocrates*. Oxford: Blackwell Publishing.

Chamberlain, J. M. (2009). Portfolio-based appraisal: Superficial or useful? *British Journal of Hospital Medicine, 70*(11), 176–179.

Conrad, P. (1992). Medicalization and social control. *Annual Review of Sociology, 18*, 209–232.

Foucault, M. (1972). *The archaeology of knowledge*. New York: Pantheon Books.

Foucault, M. (1989). *The birth of the clinic: An archaeology of medical perception*. London: Routledge Classics.
Freidson, E. (1970). *The profession of medicine*. New York: Dodds Mead.
Giddens, A. (1990). *The consequences of modernity*. Cambridge: Polity.
Giddens, A. (1991). *Modernity and self-identity: Self and society in late modernity*. Cambridge: Polity.
Huby, P. (1969). *Greek ethics*. London: Macmillan.
Illich, I. (1976). *The limits of medicine*. London: Marion Boyars.
Jewson, N. D. (1974). Medical knowledge and the patronage system in eighteenth century England. *Sociology, 8*, 369–385.
Jewson, N. D. (1976). The disappearance of the sick-man from the medical cosmology. *Sociology, 10*, 225–244.
Lindeman, M. (1999). *Medicine and society in early modern Europe*. Cambridge: Cambridge University Press.
Lupton, D. (2011). *Medicine as culture* (3rd ed.). London: Sage Publications.
McKeown, T. (1979). *The role of medicine: Dream, mirage or nemesis?* Oxford: Blackwell.
Mythen, G. (2004). *Ulrich beck: A critical introduction to the risk society*. London: Pluto.
Navarro, V. (1980). Work ideology and medicine international. *Journal of Health Services, 10*, 523–550.
Nettleton, S. (2006). *The sociology of health and illness* (2nd ed.). London: Polity Press.
Oakley, A. (1984). *The captured womb: A history of the medical care of pregnant women*. Oxford: Blackwell.
Philips, N., & Hardy, C. (2002). *Discourse analysis: Investigating processes of social construction*. London: Sage Publications.
Porter, R. (1997). *The greatest benefit to mankind: A medical history of humanity*. London: Fontana Press.
Richardson, R. (1988). *Death, dissection and the destitute*. London: Penguin.
Rose, N. (1997). *Lifelines: Biology, freedom and determinism*. Harmondsworth: Penguin.
Sigerist, H. E. (1951). The social history of medicine. *The Western Journal of Surgery, 8*, 127–137.
Sinclair, S. (1997). *Making doctors: An institutional apprenticeship*. London: Berg.
Siraisis, N. (1990). *Medieval and early renaissance medicine: An introduction to knowledge and practice*. Chicago: University of Chicago Press.
Temkin, O. (1973). *Galenism: Rise and decline of a medical philosophy*. Ithaca: Cornell University Press.
Turner, B. S. (1995). *Medical power and social knowledge* (2nd ed.). London: Sage Publications.
Turner, B. S. (2008). *The body and society* (3rd ed.). London: Sage Publications.
Wade, D., & Halligan, P. (2004). Do biomedical models of illness make for a good healthcare system? *British Medical Journal, 329*, 1398–1401.
Williams, S. J. (2003). *Medicine and the body*. London: Sage Publications.
Zola, I. (1981). Medicine as an institution of social control. In P. Conrad & R. Kern (Eds.), *The sociology of health and illness: Critical perspectives*. New York: St Martin's Press.

Chapter 3
Doctors, Patients, Managers and the State

Abstract This chapter outlines developments in medical regulation as a result of the emergence of biomedicine. It discusses how the shift to hospital medicine led to the institutionalisation of medical autonomy and the principle of self-regulation in the form of the General Medical Council through the 1858 Medical Act. This chapter outlines key developments in medical regulation, tracing as it does so both the golden age of medical power in the first half of the twentieth century as well as how from the 1960s onwards medicine increasingly came under pressure from patient's rights and consumer movements as well as the rise of hospital management and a growing state concern with efficiency and risk containment. In doing so, this chapter notes how both these external pressures for reform utilised a growing routinisation and standardisations of medical work, via the emergence of evidence-based medicine and the clinical protocol, as a mechanism from which to place medical practitioners under greater third-party managerial surveillance and control as well as to promote patient choice. This state of affairs occurred against the background of the contemporary risk-adverse governing conditions discussed in Chap. 2. This chapter outlines current developments in medical regulation, notably the introduction of revalidation as well as changes in how fitness to practise cases are managed. This chapter concludes by highlighting both the limitations of contemporary surveillance and control of medical expertise as well as how the profession/state relationship requires further attention. This sets the scene for discussion in Chaps. 4 and 5 of sociological perspectives which are concerned with analysing medical regulation. End-of-chapter self-study tasks are provided so the reader can engage in further study in relation to chapter contents.

Introduction

Having outlined the emergence of biomedicine in Chap. 2, it is necessary to next examine how during the nineteenth century the medical profession utilised this resource to establish professional privilege as well as trace key developments in the

J.M. Chamberlain, *The Sociology of Medical Regulation: An Introduction*,
DOI 10.1007/978-94-007-4896-5_3, © Springer Science+Business Media Dordrecht 2013

governance of medical expertise afterwards to bring our narrative account of the history of modern medicine up to date. Hence, in this chapter, we will focus on the professionalisation of medicine in the United Kingdom and how the regulation of the profession has operated and developed over the last 150 years since the principle of professional self-regulation was institutionalised in the form of the General Medical Council (GMC) under the 1858 Medical Act. This will set the scene for discussion of sociological perspectives pertaining to medical governance in Chap. 4 and the analysis of contemporary developments in medical governance, both in the United Kingdom and internationally, in Chaps. 5 and 6. It is important to begin by recognising that medicine could not in the strictest terms be classified as a profession prior to the 1858 Medical Act. Because professionalisation is defined, at least by medical sociologists, as the process by which an occupation seeks to obtain the status of 'profession' and in doing so engender the social and economic rewards that accompany this. Key to this process is the establishment of self-governing institutions. These must be recognised by the state and enshrined in legal statuette as independent self-regulating bodies, and as such set the standards by which new members enter the profession as well as by which professional practice is judged. Within the sociology of the professions and medical sociology, literature medicine and law have long been held to be two occupations which best illustrate professionalisation (McDonald 1995).

The concept of professionalisation is most associated with the neo-Weberian social closure perspective which is discussed in greater detail in Chap. 4. The social closure model has dominated the sociological literature for the last four decades, at least in terms of explaining collective bargaining and market control activity on behalf of certain esteemed occupations such as medicine. But a key problem this approach faces is that it tends to overemphasise the degree of autonomy and freedom from outside interference professional practitioners in reality collectively possess via their self-governing educational and regulatory institutions. This is particularly the case in regard to medical regulation today. Indeed, in this chapter, we will explore an important player in the governance of the medical profession, namely, the state. As this chapter will illustrate, the governance of professions such as medicine arguably reflects the larger governing conditions of a particular point in history. Hence, the closed shop club governance style of regulation which epitomised professional regulation for so long has in recent times given way to a more networked model based on principles of openness, transparency and accountability.

The state has intervened to reform both medicine's self-governing bodies and established statutory bodies to monitor their activity. While the importance of patients' rights movement and the introduction of managerial principles in the National Health Service (NHS) must also be accounted for as part of this process. Not least of all because they each have been, albeit in their own distinctive ways, heavily influential in causing the state to act to intervene in medical regulation. In short, as this chapter will outline, over the last 150 years, a 'light-touch' minimal state model of professional regulation has been replaced by a more 'surveillance heavy' interventionist state model. It is the nature and consequences of this shift which will be explored in subsequent chapters. But before such developments can be outlined and discussed in more detail, and as a prelude to this exploration, it is first necessary to return to the emergence of biomedicine and the process by which medical autonomy became institutionalised in the form of the GMC.

From Bedside Medicine to Hospital Medicine

> Hospital medicine was clearly an important revolution in medical thinking. Also known as the Clinic, pathological medicine, Western medicine and biomedicine, it has survived and extended itself over the last two centuries to become the dominant model of medicine in the modern world.
>
> Armstrong (1995: 393)

The previous chapter discussed how the emergence of biomedicine led to the establishment of hospital medicine and, in doing so, fundamentally changed the doctor-patient relationship. Prior to this, the bedside model of medical practice operated across European states. Here, the patient's history and personal situation dominated: The sick person was the patron, providing private fees, and the medical practitioner was the client. Although not necessarily prohibitive, the cost of treatment, particularly from a gentleman physician, was typically well beyond the ability of the poor and socially excluded. Those who could not afford to pay were by and large dependent on the charity of the church for medical treatment, or as was more often was the case than not, they resorted to using a mixture of home spun or regional folkway remedies, which had been passed down the generations. Alternatively, they could call on the services of a range of medicinal 'hobbyists', who provided their services for a small charge, including cooks, blacksmiths, druggists and grocers, amongst others. It is important to recognise that the early practice of the art of healing was not the preserve of a single occupational group. For example, the 1841 census showed that although 30,000 individuals declared themselves as 'doctors', only 11,000 actually appeared on approved registers (Moran and Wood 1993). These registers were run by the royal college of physicians (established 1518), the royal college of surgeons (established 1540 as the barber-surgeons and re-established as surgeons only in 1800) and the worshipful company of apothecaries (established 1617). Each has their own training programme, completion of which led to registration of new members. Together they formed the basis out of which the modern medical profession would emerge.

But as the nineteenth century began, they were by no means dominant, particularly outside of the city-dwelling upper and middle classes they by and large served (although the apothecary, as medicinal dispenser of potions and drugs, did have more contact and influence with the mass of people than their physician and surgeon cousins). Yes, they were provincial groupings for each of the three elements of the fledging medical profession based in major towns and cities throughout the country. But a mixture of quarks, charlatans and hobbyists by and large controlled rural areas. The emergence of biomedicine enabled the fledging medical profession to establish itself as the dominate authority on medical matters and extend itself into the rural communities to stamp out those individuals it did not approve of (Parry and Parry 1976). But first it would need to settle its internal affairs.

The organisation of the occupation of medicine reflected the prevailing social hierarchies of the time. At the beginning of the eighteenth century, there were three categories of medical practitioners in England. A not dissimilar division of labour operated across European countries. For it was firmly based on the Aristocratic mores of the ruling elites operating under the patronage of monarchial systems of control. So there was the learned physician, who was a gentleman consulted on their patron, while the surgical craftsman and the dispensing apothecary tradesman were called on as need be. Physicians may prescribe medicinal remedies, but as gentlemen, they certainly were not going to engage in trade and actually sell such items. As Carr-Saunders and Wilson (1933: 421) note: *'a gentleman might be rich and might even seek riches. But certain roads to the acquisition of riches were closed to him; in particular he must not seek riches through the avenue of "trade"'.*

Each occupation had its own admission requirements. Only a degree from Oxford or Cambridge would secure entry to royal college of physicians. Book learning was valued above all else. In contrast, surgeons and apothecaries learnt their trade by practical apprenticeship. This frequently lasted several years and was undertaken at the expense of the apprentice who paid for the privilege. The apprenticeship system by and large did produce competent practitioners, but there was concern that *'at its worst, if the master neglected his duties, or the pupil was idle and cared little to learn, the period of apprenticeship too often represented so much precious time wasted'* (Muirhead-Little 1932: 6). Women, although they had for centuries been involved in the care of the sick, were excluded from entry into all three occupations. A point we will return to in Chap. 4 is the feminist critique of medicine and medical regulation. But suffice to say for the moment that their exclusion from the male-only medical occupations of the period reflected the dominant cultural values of the time which regulated women to the status of secondary citizens (Porter 1997).

Changes in the organisation of medicine began to occur during the latter part of the eighteenth century as the industrial revolution and ascent of enlightenment ideals led to industrialisation and a huge increase in people living in cities. Indeed, the rapid socio-economic changes afoot led to an increase in the urban medical marketplace and a concurrent growth in unqualified shopkeepers trading in drugs and medicinal remedies in direct competition to apothecaries (Holloway 1966). These early forerunners to the modern chemist grew in popularity so quickly that *'apothecaries, the largest order of medical practitioners, began to feel themselves encroached on from below. The result was that when unqualified practice grew to sufficient proportions the apothecaries felt that something should be done'* (Newman 1957: 58).

But if anything was to happen, the apothecary knew full well that they would need the support of the royal college of physicians. As luck would have it, by the end of the first decade of the nineteenth century, the physicians felt they had begun to come under attack from continental medicine as the 'medical gaze' of early bio-medicine began to made its presence felt, particularly in Scotland and northern England. Increasingly continentally trained practitioners sought to generate an income by entering practice in the Middle and North of England, treating the middle classes and the upwardly mobile poor. For they brought with them the very continental idea of the differential fee: the wealthy paying more than the poor. This greatly upset the members of the royal college of physicians who felt that, as gentlemen, they should not compete with each other for payment like common workmen. Yet at the same time, young London-based physicians began to spend time on the continent, mainly in the French hospitals, as part of their initial clinical education, and they brought back with them new ideas for medical training and practice. Indeed, *'experience, from the dissection table and the hospital wards, flowed through the careers of multitudinous young Englishmen as they made the journey out and back....The year in France was, far from a passive period of observation, a veritable tour de main'* (Maulitz 1987: 136).

English medicine was changing, rejecting the Galen and humoral tradition and shifting toward the biomedical model. Yet this was a slow incremental form of change. What is more, it certainly was not welcomed by all. The established medical hierarchy (i.e. the physicians), some of whom clung to the Galenic humoral tradition, felt threatened. Time was needed to adjust to the changes under way. While perhaps most importantly, regardless of their occupational and personal allegiances, medical elites wanted to secure the authority of the royal colleges. After all, some sort of social order in the face of social change must be retained. So, using the plight of the apothecary as a cover, the colleges lobbied government and the result was the 1815 Apothecaries Act. This endorsed apothecary control over medical dispensation but, at the same time, maintained in law London physician control over medical training and certification. But the 1815 Act could only hold the changing tide back for so long. Provincial physicians, surgeons and a growing number of skilled surgeon-apothecaries were deeply concerned about the standard of medical education and practice in some city universities and teaching hospitals. Indeed, the ink was barely dry on the 1815 Act before some quarters of the professions – notably the younger members increasingly schooled in biomedicine – began publicly campaigning for more far-reaching change. As a result in 1842 a National Association of General Practitioners in Medicine, Surgery and Midwifery was formed with the goal of standardising the training and examination process.

But perhaps the most influential force for change was the increasing recognition by key elements of the profession – even by the dusty old gentlemen of the royal college of physicians – that the emergence of biomedicine allowed for upward social mobility. This is something even a gentleman could strive for if the conditions were right. And they were. After all, wasn't medicine a vocation, undertaken by the individual as a necessary social good for the benefit of all, and didn't this new emerging form of medicine show this to be the case? It was certainly rapidly becoming apparent to everybody – including patients themselves – that the emergence of biomedicine was dramatically changing the nature of the doctor-patient relationship. As a result, many practitioners felt that utilising this resource to establish a united medical profession, whose members were self-governing and equal in the eyes of the law, would be of benefit to everyone. The state, in turn, was eager to ensure its growing number of hospitals, which greatly enhanced its ability to survey and monitor the population, were properly staffed by doctors who had been appropriately trained to an acceptable standard.

By the middle of the nineteenth century, the benefits of the biomedical model and the new hospitals as population surveillance and governing tools were then readily apparent to both governing and medical elites. Medical practitioners sought to ensure they gained social and economic reward for their service to the masses. So to secure the control of the royal colleges over this new medical order, government was again lobbied and the 1858 Medical Act subsequently passed by parliament. This created the modern medical profession which still exists to this day. The Act was welcomed by medical practitioners as a positive development in the securing of medical control over both urban and rural medical training and practice while at the same time extending the influence of the fledging profession into the political sphere as an independent advisor to both the state and the population on health matters. As an editorial in the medical periodical, *Lancet* reported after the enactment of the 1858 Medical Act, *'Medicine in this country has, both in regards to Science and Polity entered into a new era...with our free institution there is scarcely a limit to the influence which the profession may come to exert in the state now that it acquired a collective and political existence'* (Lancet editorial 1858: 148).

The 1858 Medical Act and the Golden Age of Medical Power

> In 1858 the GMC was effectively a gentlemen's club. Its promise that the public could trust those it registered amounted to ensuring that there were no 'bounders' in the medical fraternity [sic] who would do dastardly things such as no gentleman would do....
>
> Stacey (1992: 204)

The 1858 Medical Act established the General Council of Medical Education and Registration (subsequently shortened to the General Medical Council and abbreviated to the GMC). The GMC was made responsible to parliament via the Privy Council, to which it had to produce an annual account of activity, but on a day-to-day level, it was autonomous in matters of the standards by which to admit practitioners, judge their performance, as well as remove them from the medical register if need be. The Act created a regulatory bargain between the profession and the state. Medicine gained the privilege of professional self-regulation in return for promising the competence of registered medical practitioners could be trusted. Medicine's altruistic principles and close association with science undoubtedly led to it being granted the privilege of professional self-regulation. However, particular historical social and political circumstances also shaped the nature of the institutional arrangements surrounding the establishment of the GMC. Moran (2004: 28) notes that *'because government was the product of an era of oligarchy, deference and social elitism it was the government of clubs...[and] the government of doctors was patterned on the club system'*. Similarly, Marquand (1988: 178) says of the ideology of the governing style of the period that *'[the] atmosphere of British government was that of a club, whose members trusted each other to observe the spirit of the club rules, the notion that the principles underlying the rules should be clearly defined and publicly proclaimed was profoundly alien'*.

The 1858 Act and the GMC were then reflections of the essentially pre-democratic, oligarchic, political structure in which they were founded. Furthermore, these were masculine affairs, women were excluded, with the door to the club being jealously guarded to prevent unwanted individuals from entering. This state of affairs was to remain unchecked for the next 100 years, although some women did manage to start to gain access as the twentieth century progressed. But old habits die hard. For example, Stacey (1992: 204–205) noted when she joined the GMC as a lay member in the 1970s that it still retrained the air of being an exclusive 'gentlemen's club': *'One felt that change was accepted reluctantly and that tradition dominated. It was really a place for white men...The few women were tolerated and treated very civilly (albeit their toilets were in basement or attic) but the ethos was male...Life on the Council was not entirely nineteenth century of course; the founding fathers would have felt out of place in a number of ways. But given their pervasive legacy, they would have felt happier there than in many parts of the outside world'.*

In part, the reason why medicine's club mentality lasted as long as it did into the latter half of the twentieth century was because this was a reflection of the individualistic self-image of themselves *'as autonomous, self-sufficient practitioners with personal responsibility for their patients'* (Davies 2004: 59). For many, the calling of medicine as a vocation meant that voluntary compliance and self-regulation felt somehow to be right. After all, they could be trusted to always put their patients first, couldn't they? Hence, informal and unwritten gentleman's agreements which emphasised self-discipline were the glue that held the club together (Baggot 2002). Exercising direct control over fellow members, essentially through interfering with their practice and questioning their professional judgement, was seen as distasteful and, indeed, felt to be largely unnecessary. For relatively few 'bad apples' were expected to exist. After all, how could such people get into the club in the first place?

It is perhaps surprising to learn that medicine's club governance style of self-regulation went largely unchecked for the next 100 years but this is what did happen. Of course, in the intervening period, incremental changes did occur to both the structure of the GMC and the accredited medical training programmes it oversees. Yet the changes relating to medical education were not about challenging the role of the GMC as such. Rather they were about ensuring practitioners were adequately prepared for their job role and trained in up-to-date developments in medical science. So an Act in 1886 was introduced to ensure new doctors had passed clinical examinations in midwifery. While an Act in 1950 extend medical training by adding an additional practice-based year (called the preregistration year) onto university-based 5-year undergraduate degree courses. An Act in 1968 formally established higher specialist training under the aegis of the royal colleges and the GMC.

Given rapid advancements in medical science and technology, the 1968 Act also made it clear that the university-based medical education period was an introduction to medical practice. Medical advancements meant that universities could not be expected to produce a fully prepared doctor. In addition to these relatively small changes as regards medical education, the structure and membership of the GMC slightly changed. So the 1886 Act allowed five members of the GMC to be elected by 'rank and file' doctors as it was felt that this change would make the GMC more representative of the profession as a whole. While in response to workload demands by the mid-1930s, the original 24 GMC board members had increased to 42. By the early 1940s, for the first time, one lay member was nominated by the Privy Council to sit on the GMC board. This was usually a retired Member of Parliament. The 1950 Act increased the size of the GMC to 50 and the lay membership to 3. There were no female members until 1951; only one female council member prior to 1971. This was in spite of the fact that by 1970, they were 12,596 women on the medical register (Stacey 1992).

In addition to these developments, the profession was in a good position when the NHS was established in 1948 as only practitioners on the GMC register could work in the NHS (Klein 1989). The state was heavily dependent on the medical profession to provide universal health care, and the profession used this opportunity to reinforce their right to clinical autonomy in the workplace as well as to self-regulate educational and disciplinary activities. So much so that the 1944 White Paper on the creation of a nationalised health service stated that *' the doctors taking part must remain free to direct their clinical knowledge and personal skill for the benefit of their patients in the way they feel best'* (Ministry of Health 1944: 26). This was the golden age of medical power and autonomy. As Elston (1997: 67) notes, medicine's *'freedom extended to include a considerable level of representation as of right on policy making bodies at all levels as well as freedom from managerial supervision over patient care'*. But medical dominance could not last forever. Indeed, the end was close at hand.

Challenging Medicine

As usual when it comes to politics, the beginning and end lay in the dual realms of ideology and economics. During the immediate post-Second World War era of consensus politics, people generally possessed a sense of deference to governing elites. But from the 1960s onwards, social protest, based around an array of issues including civil rights, women's movements, worker rights, anti-war and political reform campaigners, signified the beginning of a radical breaking with the past. People were becoming less and less willing to accept the authority of traditional somewhat elitist governing regimes as well as the wide array of experts who advised them on matters ranging from public health and social welfare to urban renewal and the design of cityscapes. At the same time, it was becoming apparent that public expenditure was a very real issue, and both political parties introduced reviews of public services. Simply put, the welfare state was too costly, and of particular concern was the NHS.

The 1979 conservative administration, led by Margaret Thatcher, was committed to the neoliberal notion that the discipline of the market and the power of consumer choice could enhance public services and reduce costs. Doctors, with their entrenched power within the NHS, were seen as a particular problem to containing costs and improve efficiency. So a review was instigated, chaired by Sir Roy Griffiths, who was the Managing Director of Sainsbury's Supermarket Chain. In 1983, the inquiry gave its recommendations. Perhaps the key recommendation was the introduction of a managerial layer to oversee the use of resources. Medical autonomy was under threat. Further NHS reforms initiated by conservative administrations throughout the 1980s and early 1990s – such as Working with Patients (Department of Health 1989a) and the Patients Charter (Department of Health 1991a) – challenged medical autonomy while seeking to reduce costs, improve efficiency, as well as make sure the voices of patients were heard. At the same time as NHS reform was underway, public suspicion on the principle of professional self-regulation was growing.

From the early 1980s onwards, a growing number of medical malpractice cases came to the foreground, many of which were televised via investigative reporter programmes such as Dispatches, That's Life, World in Action and File on Four. Particular focus was given to the role of the GMC in failing to ensure the doctors they investigated were competent and ethically sound. Even the most sympathetic critic could not help but admit that the GMC seemed out of touch and failing patients. Robinson (1988:35), who herself had been a lay member of the GMC during the 1980s, argued that *'the way doctors behave after a mistake has been made causes most criticism and really brings the profession into disrepute'*. Increasing public concern with medical mistakes meant that the GMC was forced to examine doctor's clinical performance rather than more traditional disciplinary concerns, such as inappropriate relationships with patients or instances of criminality. But as Stacey (1992: 183) noted, *'[though GMC was] now prepared to look at more cases bordering on the clinical, errors in practice did not rank as seriously in the [disciplinary] committee's mind as some other offences (advertising for example)'*.

Larson (1977) argues that medicine's collective belief in the need for doctors to possess clinical autonomy, due to the specialist nature of medical expertise, means it has developed an occupational culture which in technical terms is cognitively exclusive. Such exclusivity can lead to the development of elitism in relationships with non-group members and a general reluctance on behalf of group members to question each other's dealings with non-group members in case it is seen as breaking ranks. The GMC's cognitive exclusivity had led it to make the mistake of presuming the general public was still culturally bound to accept medical authority without question. While when it recognised that it had to change, its exclusivity led it to do so reluctantly. But with the rapidly changing and expanding nature of medical knowledge meant by the end of the 1980s, the GMC had to look at the central issue of doctors' continued competence to practise. Not least of all because they were coming under pressure to reform because of a substantial rise in medical litigation in the NHS (Allsop and Mulcahy 1996). This move by the GMC towards looking at doctors' continued competence to practise was progressive. Nevertheless, throughout the 1980s and into the early 1990s, it remained an essentially reactive institution, providing little effective leadership to the profession at large. It was representing doctors, not regulating them as it should have been, and consequently was perceived by many critical commentators to be failing as a regulatory body in its statutory duty to protect the general public (Gladstone 2000). As Slater (2000: 7) notes, *'if the profession does not fulfill its part of the bargain, then the state is obliged to reform medical regulation in order to restore public confidence'*. This is exactly what was going to happen next.

By the beginning of the 1990s, the GMC was under pressure to modernise itself. Yet the principle of medical self-regulation had not been directly challenged. But the same cannot be said for the principle of clinical autonomy in the NHS. By the mid-1980s, it was clear that managerial control over medical work was growing. So much so that by the early 1990s, Flynn noted that there had been a *'tendency during the last decade...towards an erosion of professional dominance in the face of increased...managerial power'* (Flynn 1992: 50). Indeed the 1990s began with the tacit agreement that it was time *'to develop a comprehensive set of measures of the outcome of much of the work of...doctors'* (Department of Health 1989b: 2). The state, via hospital management, had decided, *'to consider how the quality of medical care can best be improved by means of medical audit, and on the development of indicators of clinical outcome'* (Department of Health 1989b:2).

Medical audit involved embedding evidence-based medicine within the management of health-care delivery. It was used to refocus clinical practice on the use of evidence-based indicators concerning the risk management of clinical practice. Evidence-based medicine was first developed to address regional variations in key performance outcome areas, such as mortality rates following surgery and length of stay in hospital following admission (Berg 1997). By the end of the 1980s, the technological and communication revolution led to a rapid increase in the ability to survey and statistically deconstruct clinical outcomes in order to identify best-practice guidelines and protocols (Wennberg 1988). As Wennberg (1988: 34) noted, *'[it] is now possible to speak of a new set of disciplines which together constitute the evaluative clinical sciences. They offer the promise of a scientific programme that can greatly improve clinical decision-making by decreasing uncertainty about the probabilities and the value to patients of the outcomes of care. They also offer new ways of communicating information to physicians and patients that can greatly increase understanding about the consequences of medical choices and thus help patients make decisions they truly want'*. This state of affairs brought with it the forging of a new relationship between the four main stakeholders involved in health care – the state, the public, the managers and the professionals themselves. After all, the routinisation of medicine and establishment of best-evidenced clinical guidance enabled, at least in theory, the demarcatory lines between patients and managers and members of the profession to be redrawn. Yes, the tacit foundations of clinical expertise may be irreducible and unknowable to medical outsiders, but nevertheless, much everyday medical work could now be readily understood, particularly in terms of the probabilistic outcomes of different lines of action.

Indeed, the election in 1997 of the New Labour government intensified this process. A comprehensive, management-led system of clinical governance was proposed to set and monitor standards governing medical work. Clinical governance was defined as *'a framework through which the NHS organizations are accountable for continuously improving the quality of their services and safeguarding high standards of care by creating an environment in which excellence in clinical care will flourish'* (Department of Health 1998: 33). Clinical standards were set nationally by the

National Institute for Clinical Excellence (NICE). NICE makes recommendations on the cost-effectiveness of specific treatments and disseminates clinical standards and guidelines. Upon reviewing the reforms to NHS management over the previous decade, Light (1998: 431–432) stated that *'the national framework for performance management is extensive. The White Papers propose establishment of evidence-based patterns and levels of service, clinical guidelines, and clinical performance review, in order to ensure patients of high uniform quality throughout the service'*.

This was a significant challenge to the principle of professional self-regulation. Indeed, New Labour also established the Council for Healthcare Regulatory Excellence to oversee the professional bodies responsible for maintaining standards in health care (Department of Health 2001a, b, c, d). The nine professional self-regulatory councils under its remit included General Chiropractic Council (GCC), General Dental Council (GDC), General Medical Council (GMC), General Optical Council (GOC), General Osteopathic Council (GOsC), Health Professions Council (HPC), Nursing and Midwifery Council (NMC), Pharmaceutical Society of Northern Ireland (PSNI) and Royal Pharmaceutical Society of Great Britain (RPSGB). Yet the principle of medical self-regulation did remain. In part this is because the GMC is a self-funded organisation and does mean that the profession at large carries the financial burden of self-regulation. But it was equally made clear to the GMC that it had to reform itself. Sure enough, in 2003 its executive membership was reduced to 35 members, 14 (40%) of whom are lay members. However, it was also noted that *'elected and appointed medical members still form a majority on the council and the president, a doctor, is elected by the membership thus preserving a symbol of self-regulation'* (Allsop 2006:629). But then the Shipman case happened and everything changed forever.

The Shipman Case

Before the Shipman case is discussed, it is necessary to mention Bristol. Indeed, the Bristol case in the mid-1990s is often quite rightly held to be a key development in the regulation of the medical profession. It involved the deaths of 29 children following surgery at Bristol Royal Infirmary. The doctors involved had tried to cover over their clinical incompetence. Their incompetence only came to light after a colleague broke ranks and reported the cover up to the local press. The resulting public inquiry chaired by Professor Ian Kennedy reported extensively on the failings within the management and clinical systems of the NHS to identify efficiently and effectively poor clinical performance. A key part of this, it reported, was that NHS employees who had concerns with poorly performing colleagues must feel able to report them (Bristol Royal Infirmary Inquiry 2001). The Kennedy report criticised the hierarchical medical club culture present in the Bristol children's unit. It noted that this was a reflection of the wider system of professional self-regulation within medicine and other health-care professions as well.

The Bristol inquiry report established the already mentioned Council for Healthcare Regulatory Excellence which was empowered by the state to harmonise the work of health-care regulators. It was finally clear to the GMC and the royal colleges that they had to act to remove the club culture of mutual protectionism present within medicine at large. As Stacey (2000: 39) noted, *'In Britain today the balance has shifted a bit towards external governmental control, through the NHS reforms and new institutions such as NICE, but that is not all. The medical profession now seems intent upon regulating pro-actively. Both profession and government are paying more attention to the local level and its relation to national bodies such as the GMC and the Royal Colleges'*. A regular test of clinician competence, called revalidation, was proposed to ensure the fitness to practise of doctors as a condition of their staying on the medical register. It was also proposed that they undergo annual appraisals of their work performance. Annual appraisal was introduced to support doctors to maintain medical excellence within the framework of clinical governance (Department of Health 2000a). It was finally introduced nationally in 2003 as a formative developmental educational exercise (Chamberlain 2009a). In other words, though in principle and practice open to managerial input and review, annual appraisal was essentially doctor controlled. Revalidation was seen as more of a threat to medical autonomy than annual appraisal. After all, if a doctor failed the revalidation test, then they could be stuck off the medial register.

A state-sponsored consultation process addressed the issue of what final form it should take. Patient interest groups wanted revalidation to be heavily based around assessment of actual clinical performance and voiced the need for it. Conversely, rank and file members of the profession and members of its elite institutions, such as the royal colleges, under no circumstances wanted a periodic exam to form the basis for revalidation (Irvine 2001). But medical elites could see the tide was turning against them and eventually they gave in. In 1999, it was decided that *'to maintain their registration, all doctors must be able to demonstrate regularly that they continue to be fit to practice in their chosen specialty'* (GMC 1999: 1). What had to be decided would be the exact form that revalidation would take. The debate continued on and on. But then the Shipman case hit the headlines, and nothing would ever be the same again.

A general practitioner in Hyde, Greater Manchester, Dr Shipman, was a popular doctor. Between 1995 and 1998, he murdered 15 elderly patients with lethal doses of diamorphine. It was later discovered that between 1974 and 1998, he had murdered 215 patients. Doubts remained about a further 45 (Smith 2005). Soon it was discovered by the media that Shipman had previously been before the GMC's disciplinary committee in 1976 for dishonestly obtaining drugs and forging NHS prescriptions. He had been dealt with leniently and essentially let off with a warning. This, along with the fact that the GMC refused to strike him off the medical register until after his court case, signalled the start of another period of intense criticism for the GMC. The case for reform was unanswerable. The Secretary of State released a statement saying that the *'GMC…must be truly accountable and it must be guided at all times by the welfare and safety of patients. We owe it to the relatives of Shipman's victims to prevent a repetition of what happened in Hyde'* (quoted in Gladstone 2000: 10).

A public inquiry into the Shipman case, chaired by Dame Janet Smith, began with the remit of conducting far-reaching review of the GMC. During the Shipman Inquiry, the GMC made changes to its membership as it sought to respond to growing criticism of its operation.

In 2003, the GMC's executive membership was reduced from 104 to 35 members, 14 (40%) of whom were lay members. But Smith was unimpressed. She was highly critical of the GMC, annual appraisal and the plans for revalidation. She felt doctor-led annual appraisal *did 'not offer the public protection from underperforming doctors'* (Smith 2005: 1048). She also criticised how revalidation was been handled. She argued that the possibility of summative pass/fail testing had been dropped in favour of a light-touch approach to revalidation that essentially involved rubber stamping existing annual appraisals. Indeed, Smith (2005: 1174) said that the original *'proposals were unpopular with a powerful section of the profession. So the GMC retreated from its earlier vision and devised a system that it calls revalidation but which does not involve any evaluation of a doctor's fitness to practice'*. Furthermore, she actively criticised Catto's comparison of revalidation to an MOT during interviews with the media. She said, *'He [Sir Catto] expressed pride in the fact that no other country in the world had a system of time-limited license dependent upon doctors demonstrating they are up to date and fit to practice. To call revalidation an MOT for doctors is a catchword. It is easy for the listener to remember. I think that many people who heard that programme would have taken away the impression that revalidation is a test for doctors, just like the MOT. That is not a true impression'* (Smith 2005: 1086).

Smith was highly critical of the working culture of the GMC. She felt that although the GMC had changed, it had not changed enough, *'I would like to believe that the GMCs working culture would continue to change in the right direction by virtue of its own momentum. However, I do not feel confident it will do so. I am sure they are many people within the GMC, both members and staff, who want to see the regulation of the medical profession based upon the principles of 'patient centred' medicine and public protection. The problem seems to be that, when specific issues arise, opposing views are taken, and as in the past, the balance sometimes tips in the interests of doctors'* (Smith 2005: 1176). Finally, Smith (2005: 1176) argued that the elected nature of medical members on the GMC made the central issue of protecting the interests of the public difficult for members, *'it seems....that one of the fundamental problems facing the GMC is the perception, shared by many doctors, that it is supposed to be 'representing' them. It is not, it is regulating them....In fact the medical profession has a very effective representative body in the BMA, it does not need – and should not have – two'*. She concluded that she was *'driven to the conclusion that, for the majority of GMC members, the old culture of protecting the interests of doctors lingers on'* (Smith 2005: 1174). She recommended the GMC be further reviewed and its membership changed and elected members replaced with nominated members. These individuals were to be independently selected via the Public Appointees Committee on the basis of their ability to serve the public interest.

The Donaldson Report

Although it covered some key issues specifically to do with the GMC, the Smith's review of the Shipman case did not include a full and detailed inspection of GMC activity. But such a review of the GMC was undoubtedly needed. The criticisms regarding the GMC's working culture and proposals for revalidation meant the state had no choice to step in and undertake a full review of medical regulation. The task of undertaking this review fell to Sir Liam Donaldson, who at the time was the Chief Medical Officer for England. His subsequent report was published in July 2006 (Donaldson 2006) and informed the content of the 2008 Health and Social Care Act, which will be discussed shortly. Donaldson's report was extensive in its recommendations, but these can be themed down into four key points. First, it was recommended that the GMC face yearly questions from a committee of MPs and that, as had been recommended by Smith (2005), members should be elected independently via the Public Appointments Commission instead of the medical profession. Second, it was recommended that the GMC lose control over undergraduate medical education. This being taken over by the relatively newly constituted Postgraduate Medical Education Training Board (PMETB). The PMETB board has a membership of 25: 17 medical members and 8 lay members. This body had been created to oversee a new 2-year foundation-training programme for junior doctors, formally the preregistration and house officer years (BMA 2005). The Foundation Programme is an outcome-focused competence-based curriculum, possessing explicit standards and structured supervision and assessment tools. It was designed to tackle the prevalence for poor supervision and ad hoc assessments present in traditional postgraduate training, as highlighted by Sinclair (1997). The implementation of the Foundation Programme under the banner of Modernizing Medical Careers (MMC) formed part of broader reforms to higher specialty and general practice training programmes. As part of this, Donaldson recommended that clear national performance standards in each specialty be developed.

Third, Donaldson proposed that although the GMC will still investigate complaints, it would no longer make a final decision on guilt. This will be left up to an independent tribunal. Furthermore, it was recommended that the burden of proof required in fitness to practise cases will also be lessened from criminal standards, beyond all reasonable doubt, to civil standards on the balance of probability. This was recommended by Smith to ensure the public interest. Complaints will initially be dealt with at a local level by a GMC affiliate who will be appointed in each hospital and primary care trust with the most serious cases being passed up to the GMC to investigate and present the case to the tribunal.

Fourth, Donaldson proposed this local GMC affiliate should also be responsible for the first of what is a proposed two-strand version of revalidation. The first strand of revalidation, called relicensing, was to involve the local GMC affiliate via annual appraisal testing a doctor's fitness to practise so they can stay on the register of approved practitioners. Donaldson proposed that the second strand of revalidation, called recertification, should be managed by the royal colleges and involve the direct hands-on testing a doctor's fitness to practise so they can first join and subsequently remain on the specialist register. He also recommended that NHS appraisal be standardised and audited. Although Donaldson recommended NHS appraisal be separate from revalidation, he argued that like revalidation, appraisal should involve the collection of 360-degree feedback on a doctor's fitness to practise (i.e. feedback from medical and non-medical staff as well as patients).

As perhaps would be expected, Donaldson's proposals caused quite a stir amongst the medical profession. Irvine (2006:966) commented that *'the combined effect of Donaldson's measures could be quite profound. They should result in much stronger standards based; professional self-regulation led by a revitalised GMC and the royal colleges. That would be reassuring to the public and patients, strengthen doctor's professionalism, and appeal to the huge majority of conscientious doctors who take pride in the standing of their profession'*. However, others were more cautious (General Medical Council 2000). Many doctors did not think it was a positive step for the GMC to lose its powers of adjudication in relation to fitness to practise cases where a doctor can be stopped from practicing medicine through being removed from the medical register (Chamberlain 2009b). Indeed, the royal college of obstetricians and gynaecologists (2006), the academy of royal medical colleges (2006) and the royal college of general practitioners (2006a, b) all strongly opposed this idea. Neither did many in the profession like the idea of fitness to practise cases being judged on a standard less than beyond reasonable doubt. Also, in terms of GMC membership, there was a fear that the elimination of elected professional members from the GMC will result in the erosion of professionally led regulation (Kmietowicz 2006). After all, this raised the very real possibility that freely elected individuals may seek to pursue their own reform agendas. Yet everyone broadly accepted the need for revalidation (Bruce 2007). It was seen as a necessary step to take by the profession, and the proposal to strengthen the college's role in it was welcomed by both them and the GMC.

The 2008 Health and Social Care Act and Beyond

There can be no doubt Donaldson's proposals reinforced to both members of the medical profession and non-medical observers that there had been a shift towards emphasising professional accountability over professional autonomy. The changes proposed were designed to engender public confidence and trust in the service provided through the proactive management of medical risk. Yet alterations and amendments to Donaldson's proposals occurred almost immediately after their publication. Indeed, by the time the government published its own discussion paper in light of Donaldson's proposals, it had already conceded, after lobbying by the GMC, that the GMC should retain control over undergraduate medical education. While it was also agreed that GMC should take over the standard setting and quality assurance role of PMETB to streamline the regulatory bodies and keep them under one roof (GMC 2008). It was agreed that the GMC would work with the royal colleges to introduce revalidation. Consequently, medical control over entry onto (via medical school and junior doctor training) and exit from (via performance appraisal of their continued competence) the legally underwritten state approved register of practitioners is to continue.

Yet it is important to highlight that Donaldson's proposals did inform the Health and Social Care Act of 2008, and this Act significantly altered the GMC through requiring non-medical lay members to make up half of the GMC membership, with members being appointed via the Public Appointments Commission. Importantly, the grounds on which fitness to practise cases are judged was also changed. As already noted, such cases have traditionally been judged on the criminal standard: beyond all reasonable doubt. A situation that frequently led commentators to argue the GMC's disciplinary procedures first and foremost protected doctors (Allsop 2006). But the Act required that such cases now be judged on the civil standard of proof: on the balance of probability. It was argued that this would enable underperforming doctors to be more easily stopped from practicing medicine. While to enhance impartiality and the independence of the hearing process, the Act also required cases be heard by an independent adjudicator, not by members of the GMC (Department of Health 2009). Here, the Act introduced what it called a GMC affiliate. This person was to operate at a local NHS level to coordinate the investigation of patient complaints. While what is termed a responsible officer was tasked to work with NHS management, the GMC and the royal colleges to implement, at a local level, new arrangements for ensuring every doctor is fit to practise in their chosen specialty: revalidation.

Since the Bristol case, doctors had undergone an annual developmental check of their performance as part of the conditions of their NHS employment contract (Black 2002). But as already noted, Smith (2005) amongst others had felt that this process would not have flagged up Shipman as a risk to patients and did not offer the public protection from underperforming doctors. Donaldson had also agreed with this. As a result, the Act made it compulsory for doctors to pass revalidation to stay on the medical register (The Secretary of State for Health 2007). Although its exact form was held to be a matter for further consultation, it was felt that the revalidation process should involve a mixture of clinical audit, direct observation, simulated tests, knowledge tests, patient feedback and continuing professional development activates (Donaldson 2006).

In its post-Donaldson finalised guise, revalidation was going to be made up of two elements – relicensing and recertification – which incorporate NHS appraisal within them. Relicensing, it was argued, would make current NHS appraisal arrangements more rigorous, with greater direct testing of a doctor's competence in regard to key day-to-day clinical tasks. To stay on the medical register, all doctors will now have to successfully pass the relicensing requirement that they have to successfully complete five NHS annual performance appraisals. Specialist recertification was also planned to occur every 5 years. It was suggested that recertification would involve a thorough assessment of a doctor's clinical performance and be organised and quality assured by the royal college relevant to their chosen specialty. Although revalidation was originally planned for introduction in 2010, the development and piloting process has taken somewhat longer than expected. A consultation process was undertaken involving a series of what have been termed 'pathfinder projects' running between 2010 and 2011 to test out the new revalidation process in selected areas. All being well, revalidation will be introduced nationally from late 2012 to early 2013 onwards, although the process may in reality take a little longer. Importantly, as a result of the consultation and pathfinder project work conducted to date, it has been decided that revalidation will be based on local systems of more rigorously conducted annual appraisals, conducted over a 5-year period, rather than incorporating the twin processes of relicensing/recertification, and will simply affirm periodically what has already been demonstrated through the annual appraisal process. These developments will be discussed in more detail from an international perspective in Chap. 6.

The Act established the Office of Health Professions Adjudicator (OHPA) to take over the role of the GMC in the adjudication of fitness to practise cases. The stated aim of this change was to enhance impartiality and the independence of the fitness to practise hearing process within the health-care professions in general (Department of Health 2009). OHPA became a legal entity in January 2010. But in the summer of 2010, the UK government concluded that it was not persuaded of the need to introduce another regulatory body to take over the role of adjudicator in fitness to practise cases (Department of Health 2010). The governmental focus for now has moved to ensuring that GMC reform continues to enforce a shift towards a rigorous and fair complaint and fitness to practise adjudication process. Possible options voiced for consultation include a greater focus on the use of rehabilitative measures within the complaints system when concerns about a doctor's clinical performance exist, alongside the development of a more streamlined GMC tribunal system (Department of Health 2010).

In part, this decision was made in light of the stringent economic realities faced by public services in the UK as the state seeks to deal with the fall out of the 2008 global financial crisis. But it is also a reflection of the extent to which medical elites have successfully managed to subject rank and file practitioners to greater peer surveillance and control under the ever-watchful gaze of the regulatory state and its managerial imperatives (Chamberlain 2009b). In summary, what we have then is a situation where medical autonomy in its traditional club form does seem to have been successfully challenged over the last several decades. But at the same time, it is clear that for all the major changes implemented by the state to survey and monitor a seemingly more transparent and publicly accountable GMC, doctors still collectively possess significantly more autonomy and control over their regulatory, practice and educational matters, than many other occupations. What is more, they also seem to be able to exercise a certain degree of control over how measures to secure greater monitoring of a doctor's fitness to practise are implemented in practice. Subsequent chapters are concerned with exploring why this may be the case and what key consequences this state of affairs brings with it. But first, it is necessary to end this chapter with a necessary preliminary discussion of how contemporary developments in medical regulation reinforce certain core themes which must be taken into consideration we seek to examine medical regulation.

Medical Regulation and the Nature of Medical Expertise

This chapter has outlined the professionalisation of medicine and the historical development of medical regulation since the institutionalisation of the principle of professional self-regulation in the form of the GMC in 1858. After tracing the golden age of medical power and contemporary challenges to medical autonomy, this chapter outlined how the Health and Social Care Act of 2008 put into place significant checks and balances to medical control over doctor's activities. The organisation of medical regulation appears significantly different to what it was a decade ago. Let alone 150 years ago when the GMC was first established. The GMC is no longer the sole player in the medical regulatory field and now is more open and publicly accountable than it ever has been (Allsop 2006). Medical elites may argue that this state of affairs signifies that their indeed has been a cultural change towards a more transparent and contractually binding regulatory relationship between medicine and the public (i.e. Irvine 1997, 2003; Catto 2006, 2007).

Yet this chapter concluded by noting that developments since the 2008 Health and Social Care Act reinforce how medical elites possess an invaluable bargaining chip when it comes to reorganising medical regulation, namely, biomedicine and the esoteric nature of medical expertise. Both in terms of medicine's formal scientific knowledge basis and its more tacit dimensions, as individually accrued by a clinician via their exposure to the real world of messy clinical practice. Indeed, the issue of the specialist nature of professional expertise, alongside the concurrent need for professionals to exercise discretion in their work, does create a buffer zone which arguably protects doctors from outsider surveillance and control (Freidson 2001). Yes, the emergence of evidence-based medicine has brought medical autonomy under the gaze of managerial performance imperatives based around concerns with cost and risk. But as Freidson has repeatedly highlighted over an academic career spanning four decades, the need for doctors to exercise discretion in their work is an issue which is unlikely to disappear as long as people need and want to see a doctor to help them cope with illness and disease (Freidson 1970a, 1985, 1994, 2001).

Indeed, in his latest work, Freidson (2001) has moved away from his earlier more critical view of medical autonomy (i.e. Freidson 1970a). He insists that doctors must be allowed to exercise discretion in their work due to its inherently specialist nature, the tacit-indeterminate foundations of medical expertise, as well as the emphasis medicine collectively places upon providing a community service through promoting public health. He holds that non-medical external regulation of medical work is not always possible or in the public interest. He outlines three methods of regulatory control – 'bureaucracy', characterised by managerial control; 'the market', characterised by consumer control; and 'professionalism', characterised by occupational self-control (Freidson 2001). He discusses how in the last two decades greater managerially led 'bureaucracy' and a concurrent increase in the rule of 'the market' have successfully challenged 'professionalism', with the doctors increasingly losing the right to exercise discretion in their practice. In particular, he notes that patients are unwilling to adopt the subservient position medicine has historically accorded them. Patients nowadays frequently see themselves as active health-care consumers. Additionally, there has been a rise in managerial control over clinical practice through the increased use of standardised administrative procedures, in the form of clinical guidelines and protocols. These exist under the banner of supporting greater patient choice while also improving productivity. Freidson (2001: 181) argues that *'the emphasis on consumerism and managerialism has legitimized and advanced the individual pursuit of material self-interest....the very [vice] for which professions have been criticized'.*

In summary, while previously an ardent critic of the high level of autonomy granted to medicine to control its own affairs, Freidson (2001) now emphasises the positive moral role professions such as medicine can and do play in society. Like Stacey (1992, 2000) before him, he holds that the moral code of public service inherent in the concept of professionalism can act to dispel what Wilson (1990: 147) calls *'the ethical vacuum of the "postmodern" society'*. He argues that health practitioners themselves, not patients and managers, must ultimately control their work activities. Not least of all because the nature of their knowledge demands that society recognise professionals must possess *'independence of judgment and freedom of action'* (Freidson 2001: 122). Although he recognises that this may not be to everybody's taste, he calls for a revival of the 'ideology of service' and claims that professional monopolies are 'more than modes of exploitation or domination they are also social devices for supporting growth and refinement of disciplines and the quality of their practice' (Freidson 2001: 203).

Sociologists like Stacey (1992, 2000) and Freidson (1994, 2001) echo the common view amongst professionals that it is not the principle of professional self-regulation that in itself is unjustifiable. It is only particular instances where it has been abused. Professionals must now work with the public to make sure such abuses do not happen again (Irvine 2003, 2006). The advocacy by medical elites of a new professionalism, based around transparent standards and proactive performance management, is an attempt to establish a new contractual relationship between the medical profession and the public against the background of increasing government intervention into the field of medical regulation (Slater 2000). Furthermore, recent attempts to change in the field of professional regulation reinforce the fact that effective medical regulation, similar to the effective delivery of health care, requires the cooperation and proactive involvement of individual medical practitioners and their elite institutions. This is because contemporary challenges to professional autonomy bring to the foreground the fact that the principle of medical self-regulation was first institutionalised in the form of the GMC as it provided a workable solution to the complex problem of *'how to [both] nurture and control occupations with complex, esoteric knowledge and skill...which provide us with critical personal services'* (Freidson 2001: 220).

As Chamberlain (2009a) notes, often when calls for greater codification and performance management of medical work are made, these result in a feeling of disquiet within the medical profession at large with what is ultimately seen to be a politically motivated and unrealistic tendency on behalf of government to minimise clinical risk by turning medical work into a series of routine 'step-by-step' rules and procedures against which individual clinician performance can be measured. Because, for many, this fails to recognise the importance of the tacit and personal dimensions of medical expertise and the inherent risks present in messy real-world clinical practice situations. Certainly, many medical practitioners would argue that these situations are decidedly different from the sanitised world assumed by clinical guidelines and protocols. It is no wonder therefore that, regardless of their views about how it should be undertaken and by whom, many if not all doctors claim that some form of professionally led medical regulation is both necessary and in the public interest. Make no mistake, there will be no return to the closed shop era of club governance. Indeed, medical elites must now increasingly advocate a transparent and inclusive governing regime under the ever-watchful eye of the state. Nevertheless, doctors still possess significant amount of freedom to control their own affairs, particularly when compared to other occupations. The current situation concerning the governance of medical expertise is therefore best summed up by Moran (1999: 129–130) who argues that '*…states are more important than ever before, either in the direct surveillance of the profession or in supervising the institutions of surveillance… [this] has not necessarily diminished the power of doctors; but it has profoundly changed the institutional landscape upon which they have to operate*'.

This chapter has traced the historical development of the principle of medical self-regulation in the United Kingdom. The events it discussed do seem to add weight to the argument that medical autonomy has declined somewhat in the last three decades from the golden age when doctor knew best. Individually and collectively, doctors have become more accountable for their actions. Yet it was also noted that doctors still possess a significant degree of control over their regulatory affairs and day-to-day work activities. However, a key paradox surrounding recent challenges to medical autonomy in the form of a doctor's clinical freedom at the bedside and the principle of medical self-regulation is that they have occurred at a time when the success of medical knowledge and technology to promote public health is greater than it has ever been (Gabe et al. 2004). Here, it must be remembered that both this and the previous chapter have highlighted how modern medicine and the modern state are entwined entities.

Certainly, the close relationship between medicine's club mentality and the Victorian style of club governance illustrates that the development of modern medicine and the principle of medical self-regulation is interwoven with the development of the modern state. Consequently, as Moran (1999, 2004) argues, instead of signifying medicines apparent decline, it can be said that recent challenges to professional self-regulation bear witness to the fact that there has been a fundamental shift in the legitimate grounds for the practice of good governance throughout all spheres of contemporary public life. Perkin (1989) similarly argues that there has been a backlash against professional society as part of a profound shift in public attitudes towards institutional authority. This he holds coincided with the political and economic re-emergence of liberalism in the 1970s (Stacey 1992). Whether one agrees with Perkin or not, the close relationship between medicine and the state highlights the necessity of exploring how sociologists have conceptualised the governance of medical expertise. This task is the focus of Chaps. 4 and 5.

Conclusion

This chapter has focused on outlining the development of medical regulation in light of the emergence of biomedicine and how this transformed the relationships between the state, the profession and the public. This chapter outlined how as biomedicine developed during the nineteenth century the medical profession began to seek to utilise this new resource to establish control of the medical marketplace. The end result of this activity – the 1858 Medical Act – established the principle of professional self-regulation in the form of the GMC as well as state-endorsed occupational control of the medical marketplace. This placed the medical profession in an ideal position when the National Health Service was created. This engendered the golden age of medical dominance. However, this chapter went on to outline how the emergence of the patient's right movement and the rise of hospital medicine challenged institutionalised medical power. This chapter then went on to discuss the Bristol and Shipman cases and how the resulting governmental enquiries acted to make the medical profession and its regulatory institutions more open and accountable. This chapter then outlined key reforms introduced by the state via the 2008 Health and Social Care Act. Firstly, revalidation has been introduced to ensure the fitness to practise of medical practitioners. Secondly, reforms to how medical malpractice cases are handled have been introduced. Importantly, the grounds on which fitness to practise cases are judged has been changed. Such cases have traditionally been judged on the criminal standard: beyond all reasonable doubt. A situation that frequently led commentators to argue the GMC's disciplinary procedures first and foremost protected doctors. But now, such cases will be judged on the civil standard of proof: on the balance of probability. It was argued that this would enable underperforming doctors to be more easily stopped from practicing medicine. Additionally, efforts have been made to enhance the impartiality and the independence of the hearing process. Subsequent chapters will discuss these changes in more detail. For the moment, it is important to conclude that this chapter noted the close relationship that exists between the state and the medical profession and how forms of governance must be undertaken within contemporary society. This sets the scene for the next two chapters which outline how medical regulation has been conceptualised by sociologists. The end-of-chapter self-study activities will help you to consolidate what you have learnt before moving on to consider the contents of chapter 4.

Self-Study Activity

1. Write a 1,000-word essay which outlines key developments in the regulation of the medical profession since the 1858 Medical Act and critically evaluates the role of patient rights movement in challenging institutionalised medical autonomy.
2. Produce a 15-min PowerPoint presentation which critically considers the role of the evidence-based medicine in enabling the surveillance and performance

management of medical work. What other factors do you think should be used to evaluate the efficiency of a doctor's work activity? Why?

3. Write a 1,000-word essay which critically evaluates the idea that changes in medical regulation are linked to broader changes in the nature of contemporary governance which in turn are due to the emergence of the risk society.

References

Academy of Royal Medical Colleges. (2006). *Response to good doctors, safer patients*. Published by The Academy of Royal Medical Colleges.

Allsop, J. (2006). Regaining trust in medicine: Professional and state strategies. *Current Sociology, 54*(4), 621–36.

Allsop, J., & Mulcahy, L. (1996). *Regulating medical work: Formal and informal controls*. Milton Keynes: Open University Press.

Armstrong, D. (1995). The rise of surveillance medicine. *Sociology of Health & Illness, 17*, 393–404.

Baggot, R. (2002). Regulatory politics, health professionals, and the public interest. In J. Allsop & M. Saks (Eds.), *Regulating the health professions* (pp. 160–187). London: Sage Publications.

Berg, M. (1997). Problems and promises of the protocol. *Social Science & Medicine, 44*(8), 1081–88.

Black, D. (2002). *Consultant appraisal in GMC news supplement on appraisal and revalidation 1–4*. London: The GMC.

BMA. (2005). *The foundation programme: A rough guide*. London: Stationary Office.

Bristol Royal Infirmary Inquiry. (2001). *Learning from Bristol: The report of the public inquiry into children's heart surgery at the Bristol Royal Infirmary, 1984–1995*. London: Stationary Office London.

Bruce, D. A. (2007). Regulation of doctors. *British Medical Journal, 334*, 436–37.

Carr-Saunders, A. M., & Wilson, P. A. (1933). *The professions*. London: The Clarendon Press University Press.

Catto, G. (2006). *GMC news 3*. London: GMC.

Catto, G. (2007). Will we be getting good doctors and safer patients? *British Medical Journal, 334*, 450.

Chamberlain, J. M. (2009a). Portfolio-based appraisal: Superficial or useful? *British Journal of Hospital Medicine, 70*(11), 176–177.

Chamberlain, J. M. (2009b). *Doctoring medical governance: Medical self-regulation in transition*. New York: Nova.

Davies, C. (2004). Regulating the health care workforce: Next steps for research. *Journal of Health Services Research & Policy, 9*(1), 55–61.

Department of Health. (1989a). *Working for patients*. London: DOH.

Department of Health. (1989b). *Medical audit; NHS review* (Working Paper 6). London: DOH.

Department of Health. (1991a). *The patient's charter*. London: DOH.

Department of Health. (1991b). *Junior doctors: The new deal*. London: DOH.

Department of Health. (1998). *A first class service: Quality in the new NHS*. London: DOH.

Department of Health. (2000a). *The NHS plan*. London: DOH.

Department of Health. (2000b). *Patient and public involvement in the NHS*. London: DOH.

Department of Health. (2001a). *NHS appraisal: Appraisal for consultants working in the NHS*. London: DOH.

Department of Health. (2001b). *Assuring the quality of medical practice: Implementing supporting doctors, protecting patients*. London: DOH.

Department of Health. (2001c). *Modernizing regulation in the health professions*. London: DOH.

Department of Health. (2001d). *Hospital, public health medicine and community health services medical and dental staff in England 1990–2000*. London: DOH.

Department of Health. (2009). *Health care regulation: Tackling concerns nationally*. London: DOH.
Department of Health. (2010). *Fitness to practice adjudication for health professionals*. London: DOH.
Donaldson, L. (2006). *Good doctors safer patients: Proposals to strengthen the system to assure and improve the performance of doctors and to protect the safety of patients*. London: DOH.
Elston, M. A. (1997). *The sociology of medical science and technology*. Oxford: Blackwell.
Flynn, R. (1992). *Structures of control in heath management*. London: Routledge.
Freidson, E. (1970a). *The profession of medicine*. New York: Dodds Mead.
Freidson, E. (1970b). *Professional dominance: The social structure of medical care*. New York: Atherton.
Freidson, E. (1985). The reorganisation of the medical profession. *Medical Care Review, 4*(1), 1–20.
Freidson, E. (1994). *Professionalism reborn: Theory, prophecy and policy*. Cambridge: Polity Press.
Freidson, E. (2001). *Professionalism: The third logic*. Cambridge: Polity Press.
Gabe, J., Bury, M., & Elston, M. A. (2004). *Key concepts in medical sociology*. London: Sage Publications.
General Medical Council. (1999). *Report of the revalidation steering group*. London: GMC.
General Medical Council. (2000). *Revalidating doctors ensuring standards ensuring the future*. London: GMC.
General Medical Council. (2008). Bulletin Issue 42 February 2008. London: GMC.
Gladstone, D. (Ed.). (2000). *Regulating doctors*. London: Institute for the Study of Civil Society.
Holloway, T. (1966, July 10). The Apothecaries Act, 1815: A reinterpretation. *Medical History*. pp. 107–129
Irvine, D. (1997). The performance of doctors 1: Professionalism and self-regulation in a changing world. *British Medical Journal, 314*, 1540–50.
Irvine, D. (2001). Doctors in the UK: Their new professionalism and its regulatory framework. *The Lancet, 358*, 1807–10.
Irvine, D. (2003). *The doctors tale: Professionalism and the public trust*. London: Radcliffe Medical Press.
Irvine, D. (2006). Success depends upon winning hearts and minds. *British Medical Journal, 333*, 965–6.
Klein, R. E. (1989). *The politics of the national health service* (2nd ed.). London: Longman.
Kmietowicz, J. (2006). BMA rejects chief medical officers proposals on medical regulation. *British Medical Journal, 333*, 967.
Larson, M. S. (1977). *The rise of professionalism: A sociological analysis*. California: University of California Press.
Light, D. W. (1998). Managed care in a new key: Britain's strategies for the 1990s. *International Journal of Health Sciences, 28*(3), 427–44.
Marquand, D. (1988). *The unprincipled society*. London: Fontana Press.
Maulitz, R. C. (1987). *Morbid appearances: The anatomy of pathology in the early nineteenth century*. Cambridge: Cambridge University Press.
McDonald, K. M. (1995). *The sociology of the professions*. London: Sage Publications.
Ministry of Health and Department of Health for Scotland. (1944). A National Health Service CMND 6502. London: HMSO.
Moran, M. (1999). *Governing the health care state; a comparative study of the United Kingdom, the United States and Germany*. Manchester: Manchester University Press.
Moran, M. (2004). Governing doctors in the British regulatory state. In A. Gray & S. Harrison (Eds.), *Governing medicine: Theory and practice* (pp. 88–111). Milton Keynes: Open University Press.
Moran, M., & Wood, B. (1993). *States, regulation and the medical profession*. Milton Keynes: Open University Press.
Muirhead-Little, E. (1932). *The history of the British Medical Association: 1832 to 1932*. London: BMA Publishing.
Newman, C. (1957). *The evolution of medical education in the 19th century*. Oxford: Oxford University Press.

Parry, N., & Parry, J. (1976). *The rise of the medical profession: A study of collective social mobility*. London: Croom Helm Ltd.
Perkin, H. (1989). *The rise of professional society: England since 1980*. London: Routledge.
Porter, R. (1997). *The greatest benefit to mankind: A medical history of humanity from antiquity to the present*. London: McMillan.
Robinson, J. (1988). *A patients voice at the GMC: A lay member's view of the general medical council*. London: Health Rights.
Royal College of General Practitioners. (2006). *Response to good doctors, safer patients*. Published by Royal College of General Practitioners.
Royal College of Obstetricians and Gynecologists. (2006). *Response to good doctors, safer patients*. Published by The Royal College of Obstetricians and Gynecologists.
Sinclair, S. (1997). *Making doctors: An institutional apprenticeship*. London: Berg.
Slater, B. (2000). *Medical regulation and public trust*. London: An International Review Kings Fund Publishing.
Smith, J. (2005). *Shipman: Final report*. London: DOH.
Stacey, M. (1992). *Regulating British medicine*. London: Wiley.
Stacey, M. (2000). The general medical council and professional self-regulation. In D. Gladstone (Ed.), *Regulating doctors* (pp. 45–65). London: Institute for the Study of Civil Society.
The Secretary of State for Health. (2007). *Trust, assurance and safety: The regulation of health professionals in the 21st century*. London: The Stationary Office.
Wennberg, J. (1988). Practice variations and the need for outcomes research. In C. Ham (Ed.), *Health care variations: Assessing the evidence* (pp. 4–24). London: Kings Fund Institute.
Wilson, E. (1990). *Hallucinations: Life in the post-modern city*. London: Hutchinson Radius.

Chapter 4
Sociological Deconstructions I: Critiquing Medical Autonomy and Altruism

Abstract This chapter focuses on the development of the sociological study of the professions and professional regulation. It discusses how during the first part of the twentieth century functionalist sociologists focused on the altruistic ethos which arguably underpins professional occupations such as law and medicine. Here, the chapter notes how possession of a socially valued esoteric form of expertise allows certain occupational groups to obtain a legally underwritten state-endorsed monopoly over entry into an occupation and the standards by which both neophyte and more seasoned practitioners are judged. Certainly, early commentators recognised the important role played by the possession of specialist expertise in both enabling occupational control of the practitioner-client relationship and establishing state-endorsed formal self-regulatory systems for entry, training and exit from a profession. Yet they also acknowledged the key role played in legitimising occupational claims to professional status by the apparent willingness of certain groups to use their esoteric knowledge for the benefit of others. After exploring the impact of this focus on occupational altruism for the study of professional power, this chapter outlines how in the latter half of the twentieth century sociologists began to question the altruistic claims by professional groups as well as explore in more critical detail the relationship between professional occupations such as medicine and the modern state. Here, the chapter outlines the emergent neo-Weberian, neo-Marxist and feminist critiques of medical autonomy and altruism which emerged during the 1970s and 1980s. In doing so, this chapter provides a basis from which Chap. 5 traces subsequent developments in the sociological study of the professions and medical self-regulation, including, most importantly, the emergence of the Foucauldian governmentality perspective alongside the contemporary emergence of managerial and lay challenges to medicine. End-of-chapter self-study tasks are provided so the reader can engage in further study in relation to chapter contents.

J.M. Chamberlain, *The Sociology of Medical Regulation: An Introduction*,
DOI 10.1007/978-94-007-4896-5_4,

Introduction

Aside from highlighting the historically contingent nature of the emergence of the modern medical profession, Chap. 3 reinforced three important interwoven themes which are central to the study of medical regulation. First, there is the matter of medicine's esoteric expertise and how this possesses an indeterminate and deeply personal and tacit dimension while at the same time also being open to standardisation, codification and routinisation and so, in principle at least, third-party surveillance and control. Second, there is the nature of the relationship between profession and state and how in many ways they are entwined entities to the extent that, at least as forms of governance which seek to shape social interactions with others, possess a certain symbiotic relationship. Third, that the process by way of which an occupation becomes a profession is built around the possession (and successful mobilisation) of a distinctive form of specialist knowledge essential to successfully complete certain occupational tasks which cannot be wholly appropriated by others and indeed cannot be acquired by new occupational members either without some period of prolonged training and performance testing overseen by more experienced peers.

This focus on esoteric knowledge and its acquisition via apprenticeship is particularly important as it is from this that professional monopolies in the form of systems of licensing, regulation and examination develop and in doing so further exclude non-group members from the evaluation of work activity. Together these themes serve to reinforce that professionals have institutionalised expertise at the basis of their power. That is, professions such as medicine or law assert their jurisdictional claim over certain objects and problems on a day-to-day basis as a practitioner performs work tasks, but perhaps more importantly, at a broader structural level, these jurisdictional claims operate in terms of control over social organisations and institutions, such as education, research and regulatory quality assurance bodies (Abbott 1988). What is more, professional jurisdictional claims are particularly strong when they are legally underwritten by the state. In short, the knowledge gap between an occupational group and its client must be big enough to justify its authority, but it is equally important that this authority be also recognised by the state. However, it is important to add to this that the possession of a service orientation, as much as some form of esoteric expertise, is just as essential as specialist knowledge in recognising what a profession is and is not (McDonald 1995).

This brings us to an important point regarding the nature of the professional enterprise in general and the medical enterprise in particular, namely, that the development of biomedicine and the hospital clinic is undoubtedly important to the establishment of medicine as a self-regulating profession. But so is its claim to be an altruistic discipline which utilises the methodological fruits of modern science and technology for the benefit of the health of all humankind. Indeed, although early students of the professions at the beginning of the twentieth century recognised the importance of the esoteric expertise possessed by the occupational type claiming the social status 'profession', they nevertheless also focused on how certain occupational groups within society claim to possess high ethical standards and indeed rhetorically state that they place their client's welfare and interests before their own. Such explicit moral codes govern the behaviour of occupational members towards each other and society as a whole, as the famous Hippocratic Oath of 'do no harm' does in the case of medicine. As we shall discuss, this collectivity orientation was seen by early sociologists to act as a stabilising force to the excesses of the growing enterprise culture of capitalist industrial society, whose primary concern was taken to be with profit (Turner 1995).

Whether or not this viewpoint regarding capitalist society was correct, early sociologies focus on the altruistic connotations associated with the concept of professionalism reflected the concern of functionalist sociology with how social consensus and social order are maintained. It also provided the basis from which later more critical neo-Weberian, neo-Marxist and feminist standpoints emerged from the 1970s onwards. This chapter focuses on exploring these developments. In doing so, it draws together the key themes discussed in the previous chapter through the lens of examining the historical development of the sociological study of the professions and medical regulation. Specifically, it explores the sociological literature in relation to the dual concepts of profession and professional autonomy with an eye towards examining the meanings and consequences of the specialist and esoteric nature of medical expertise alongside the nature of the profession/state relationship. In doing so, it sets the scene for Chap. 5 which outlines and theorises the contemporary challenges to medical autonomy in the form of the rise of managerialism and calls by patient rights movements, amongst others, for more transparent and accountable forms of professional governance.

Functionalism: Professionalism as Altruism

We must begin our analysis of the sociology of the professions and the regulation of the medicine through the work of one of the founding fathers of functionalist sociology, Emile Durkheim. Durkheim (1957) was particularly concerned with the effects of the vast social changes which had occurred during the eighteenth and nineteenth centuries and how against this background social order could be maintained. Here, Durkheim viewed professional groups as important preconditions to the generation of social stability and consensus in society. His concern with the professions as a stabilising force to what he felt was the excessive individualism of laissez faire capitalism stems from his view of society as an organism constantly striving for equilibrium. He argued that individuals within pre-industrial societies possessed shared values and beliefs that generated a social consensus called mechanical solidarity. In short, from a young age, new members of society were socialised by family and caregivers into the norms and values of the cultural group to which they belonged, and this tied individuals together into a cohesive and functioning social order. However, he argued that traditional forms of moral authority, which generated collective norms and values, were being undermined by a growing specialisation within the division of labour. This was due to the increasingly complex nature of society as it became industrialised during the eighteenth and nineteenth centuries progressed.

Durkheim discussed how industrialisation rapidly led to the emergence of new forms of employment and the displacement of traditional social bonds as individuals moved from rural to urban areas seeking social mobility and a better standard of life. He felt that this state of affairs could cause alienation and anomie (i.e. antisocial individualism) amongst the general populace. This worried Durkheim considerably. As he believed that when collective norms and values declined, social restraints similarly decayed. This could lead to a situation where *'nothing remains but individual appetites, and since they are by nature boundless and insatiable, if there is nothing to control them, they will not be able to control themselves'* (Durkheim 1957: 11). But all was not lost. Durkheim argued that a new form of organic solidarity was emerging within modern industrialising societies. This was based upon the recognition of the need for cooperation between individuals due to their growing functional interdependence within the social sphere as society became more complex. As part of this, he held that the professions formed moral communities which promoted values such as selflessness that engendered social consensus and organic solidarity.

Whatever the weaknesses of Durkheim's viewpoint, key of which perhaps being that he possessed an overly homogenous notion of social bonds and shared culture, his arguments nevertheless informed much of the subsequent sociological analysis of the professions until the 1960s. For instance, Tawney (1921) held that the economic individualism of capitalism was inherently destructive to the community interest and that the morality of professionalism could be used to counter its excesses. He stated that *'the difference between industry as it exists today, and profession is, then, simple and unmistakable. The essence of the former is that its only criterion is the financial return, which it offers its shareholders. The essence of the later is that though men enter it for the sake of livelihood the measure of their success is the service which they perform, not the gains which they amass'* (Tawney 1921: 94–95).

Similarly, Parsons (1949) emphasised the social altruism of professional groups by arguing they possessed what he termed a collectivity orientation. That is, it was held that what set professional groups such as doctors aside from other occupations was their willingness to put their client's needs above their own wish from financial and social status reward. While Carr-Saunders and Wilson (1933: 497) held that professions *'inherent, preserve and pass on a tradition...they engender modes of life, habits of thought and standards of judgment which render them centres of resistance to crude forces which threaten steady and peaceful evolution...The family, the church and the universities, certain associations of intellectuals, and above all the great professions, stand like rocks against which the waves raised by these forces beat in vain'*.

The early functionalist hegemony regarding the sociological study of the professions also revealed itself in the work of authors who were concerned with identifying characteristics which taken together denote that an occupation is a profession. For example, Etzioni (1969) classified occupations into professions and semi-professions based upon characteristics such as length of training. While Barber (1963: 671) held that professions possessed four essential attributes – a high degree of generalised and systematic knowledge, an orientation towards the interest of the community instead of individual self-interest, a high degree of self-control exercised by practitioners over behaviour through the possession of a code of ethics internalised during a prolonged period of education and training, and finally, a reward system of monetary and status rewards that are symbolic of work achievement not self-interested gain.

To this day, occupations such as medicine protest that they possess an ideal of public service when they seek to justify collective privileges such as the principle of self-regulation and the individual social and economic rewards which come with the possession of professional status. The previous chapter discussed how contemporary changes in the governance of medical expertise have led commentators to re-emphasise the positive social role played by the professions in society (i.e. Freidson 2001). Yet the core problem with the early functionalist approach to the sociological analysis of the concept of professionalism is that it takes uncritically the altruistic claims of occupations calling themselves professions at face value, while it also views the task of sociology as being to quantify and measure the concept, 'professionalism'. Furthermore, the functionalist approach to the analysis of professionalism was criticised for being largely ahistorical. It lacked consideration of the process by which occupations utilised their cognitive and altruistic resources to exercise power in order to initially gain and subsequently maintain the social and economic rewards associated with the possession of professional status (Johnson 1972).

Sociologists were coming to realise that they were starting their analysis of the professions with the wrong question. As Hughes (1963: 656) wrote *'in my studies I passed from the false question 'Is this occupation a profession?' to the more fundamental one 'What are the circumstances in which people in an occupation attempt to turn it into a profession and themselves into professional people?'*. Hughes was highlighting that classifying an occupation as a profession was what society did, and it was not the task of sociology to do it in more scientific terms. Rather, its focus should be on investigating the socio-economic and political circumstances out of which the concept of professionalism arose. This signalled the beginning of a more critical turn in the sociological study of the professions. In contrast to the functionalist viewpoint, this focused upon the material and symbolic benefits gained from the possession of an occupational monopoly over license to practise (McDonald 1995). According to this more critical viewpoint, *'professionalism is not a set of traits which jobs have in common, nor a distinct ethic, but a mode of occupational control'* (Moran and Wood 1993: 25).

Critiquing Professional Power: The Neo-Weberian Viewpoint

The 1960s and 1970s were a time of sustained social protest and change. This was the age of the civil rights and feminist movements, calls for deep-rooted changes in how society was governed, as well as a broad rejection of traditional elitist forms of authority. Universities, via the student protest movement, were caught up in this whirlwind. It is perhaps no surprise to learn therefore that by the start of the 1970s, sociologists were turning away from the viewpoint that the professions transcended the unbridled self-interest they held to be symptomatic of modern society (McDonald 1995). As Johnson (1972: 25) wrote, *'The professional rhetoric relating to community service and altruism may be in many cases a significant factor in moulding the practices of individual professionals, but it also clearly functions as a legitimation of professional privilege'*. Functionalist sociologists mostly accepted the altruistic claims to public service espoused by professions such as medicine. Indeed, they often endorsed the fact that this separated them from other occupational groups.

However, the 1970s saw social scientists question increasingly the legitimacy of the self-espoused altruistic tendencies and value-neutral knowledge claims of occupational groups, which possessed professional status. In the context of the medical profession, they began to focus upon how medical professionalism has operated ideologically as an exclusive form of occupational control. This was seen to operate both at the micro-level of everyday interaction through the concept of clinical freedom at the bedside and the macro-institutional level through the principle of state-licensed self-regulation. They highlighted how poorly performing doctors, and in some cases even criminals, were being shielded from public accountability by the club rule of mutual protectionism inherent within medicine's self-regulatory system. This focus upon professional self-interest as opposed to professional altruism lay at the heart of the growing symbolic interactionist critique of the early functionalist view of the professions in American sociology.

The interactionist viewpoint assumes reality is socially constructed in and through everyday social interaction. Consequently, it viewed professionalism as *'an ascribed symbolic, socially negotiated status based on day-to-day interaction'* (Allsop and Saks 2002: 5). Studies of the medical profession inspired by this viewpoint, such as Becker's Boys in White (1961), highlighted that *'[the] professional principles of altruism, service and high ethical standards were...less than perfect human social constructs rather than...abstract standards which characterized a formal collectivity'* (McDonald 1995: 4). Yet, instead of focusing on the micro-individual level of the individual professional interacting within his or her work sphere, the growing critique of the professions in the Anglo-American literature primarily focused on the macro-organisational and societal level. This was largely informed by neo-Weberian sociology.

Like Durkheim, the sociologist Max Weber focused upon trying to understand emerging new social patterns in the nineteenth century caused by the rise of industrial technology, the growth of scientific knowledge and the greater potential than ever before for participation by the general populace within the political sphere. Weber was a polymath interested in law, economics, politics, science, religion as well as sociology. A key unifying theme in his writing is the idea that the progressive rationalisation of life was the main directional trend in western civilisation. By rationalisation, Weber meant a process by which explicit, abstract, calculable rules and procedures (what he called formal rationality) increasingly replaced more traditional and personal, social values and ways of life (what he called substantive rationality) at the organisational and institutional levels which govern social life (Gerth and Wright Mills 1946). Rationalisation leads to the displacement of religion by specialised rationalistic knowledge and scientific expertise. It also leads to the replacement of the skilled worker and artisan with the factory production line and machine technology. It demystifies and instrumentalities life and *'means that...there are no mysterious incalculable forces that come into play, but rather that one can, in principle, master all things by calculation'* (Weber 1947: 139).

Though Weber did not specifically address the issue of the growth of the professions, his concept of rationalisation is clearly tied to the development of modern scientific forms of expertise, of which modern medicine is a part. As Murphy (1988: 246) *notes, '[the] process of formal rationalization has generated a new type of knowledge, the systematic, codified, generalized (which implies abstract) knowledge of the means of control (of nature and of humans)'*. This is a point this chapter will return to in a moment. However, it is important to note here that sociologists with a historical bent, such as and Berlant (1975) and Larkin (1983), primarily drew upon Weber's economic theory of monopolisation when analysing the initial growth and subsequent development of professions such as medicine (Weber 1978). In doing so, they highlighted collective preoccupations with pecuniary interests, securing economic and technical domains, as well as consolidating positions of high social status and power within the sociopolitical arena. This was to be expected as Weber views professionals as a privileged commercial class, alongside bankers and merchants. He holds that they seek to exclude competitors and reap economic and social rewards through pursuing strategies that enable them to monopolise the marketplace for their services by controlling market entry and supply. By engaging in collective social mobility (i.e. the formation of group organisations and political pressure groups), occupational groups such as medicine seek to obtain privileges from the political community, to become what Weber (1978: 342) calls a legally privileged group, and ensure *'the closure of social and economic opportunities to outsiders'*.

Freidson and Social Closure

Two key early proponents of the neo-Weberian social closure model of the professions were Freidson (1970a, b) and Larson (1977). As his work came chronologically first, this chapter will discuss Freidson before moving on to Larson. In 1970, Freidson published his landmark study of the American medical profession, *Profession of Medicine*. In line with Weber's social closure perspective, Freidson held that medicine was a particularly powerful example of how professionalism operated ideologically as a form of occupational control to ensure control of the market for services. Freidson (1970a, b: 137) highlighted that the professions possessed three powerful interlocking arguments on which they justified their privileged status: *'Professional people have the special privilege of freedom from the control of outsiders. Their privilege is justified by three claims. First, the claim is that there is such an unusual degree of skill and knowledge involved in professional work that non-professionals are not equipped to evaluate or regulate it. Second, it is claimed that professionals are responsible – that they may be trusted to work conscientiously without supervision. Third, the claim is that the profession itself may be trusted to undertake the proper regulatory action on those rare occasions when an individual does not perform his work competently or ethically'.*

Freidson recognised that medical autonomy must be viewed as having limits as the state was involved in the organisation and delivery of health care. Occupations must submit to its 'protective custody' to reap the social and economic rewards associated with being a profession. Nevertheless, the state largely left doctors alone to control the technical aspects of their work. This made it for him such a good example of what a profession is. He argued that *'so long as a profession is free of the technical evaluation and control of other occupations in the division of labor, its lack of ultimate freedom from the state, and even the lack of control over the socio-economic terms of work, do not significantly change its essential character as profession'* (Freidson 1970a, b: 20). Freidson discussed how medical professionalism operated ideologically as a form of occupational control at the micro-level of everyday interaction through the concept of clinical freedom at the bedside, as well as at the macro-institutional level through the principle of state-licensed self-regulation.

The common link between the micro and macro aspects of medical autonomy for Freidson was the need for a doctor to exercise personal judgement and discretion in her work due to the inherently specialist nature of medical work (Freidson 1970a, b, 1994, 2001). This state of affairs was legitimised by the scientific basis of modern medical expertise and public acceptance of medicine's altruistic claim that it put patient need first. Furthermore, Freidson argued that medicine's freedom to control the technical evaluation of its own work had led to it possessing a high level of dominance and control not only over patients but also over the work of other health-care occupations, such as nursing. Freidson (1970a, b: 137) stated that medicine *'has the authority to direct and evaluate the work of others without in turn being subject for formal direction and evaluation by them. Paradoxically its autonomy is sustained by the dominance of its expertise in the division of labor'.*

In *Profession of Medicine* (1970a, b) and his other major study, *Professional Dominance* (1970a, b), Freidson was concerned with mapping out the negative consequences of medical autonomy in the Anglo-American context. He concluded that the dominance of medicine in the health-care arena had a negative effect on the quality of health care patients received. For Freidson, medicine was failing to self-manage satisfactorily its affairs and ensure that adequate quality control mechanisms to govern doctor's day-to-day activities were in place. Freidson (1970a, b: 370) believed that the development of unaccountable, self-governing institutions surrounding medical training and work had led to the profession of medicine to possess *'a self-deceiving vision of the objectivity and reliability of its knowledge and the virtues of its members....[Medicine's] very autonomy had led to insularity and a mistaken arrogance about its mission in the world'.*

Larson and Medical Knowledge

There can be no doubt that Freidson argued forcefully that medicine was a powerful example of how professionalism operated ideologically as a form of occupational control. For him, it was a publicly mandated state-supported supplier of a valued service, exercising autonomy in the workplace. This included dominance over other occupations in the health-care division of labour as well as collegiate control over recruitment, training and the regulation of members' conduct. Regardless of Freidson's critical insights, his work lacked a thorough historical dimension. Aware of this, Larson undertook a historical analysis of the rise of professionalism as a legitimate form of occupational control in her (McDonald 1995). She discusses how by engaging in a professional project occupations such as medicine sought to become professions by obtaining a monopoly over the market for their services and enhancing the standing of group members within the social and political spheres: *'My intention....is to examine how the occupations we call professions organized themselves to attain market power...Professionalization is thus an attempt to translate one order of scare resources – special knowledge and skills – into another – social and economic rewards. To maintain scarcity implies a tendency to monopoly: monopoly of expertise in the market, monopoly of status in the system of stratification'* (Larson 1977: xii and xvii).

As Chap. 2 noted, the rise of the clinical gaze of modern medicine in the eighteenth century changed the nature of the doctor-patient power relationship in favour of the medical profession (Jewson 1974, 1976). Bound up with this was a growing focus upon gaining direct personal experience of clinical phenomena on which to build craft expertise and justify clinical decisions. This is not to say that the increasingly formal and scientific aspects of medical expertise did not play a vitally important role in medicine's successful claim to professional status. Clearly they did. But as Larson (1977) notes, in addition to the possession of a formal knowledge base, it requires the presence of a high level of 'indetermination' in the exercise of expert judgement and technique for the monopolistic claims of the professional project to succeed. Larson (1977: 31) specifies that an occupation's knowledge base must be *'formalized or codified enough to allow standardization...and yet ...must not be so clearly codified that it does not allow a principle of exclusion [or discretion] to operate'*. Furthermore, Larson (1977: 41) observes that *'the leaders of the professional project will define the areas that are not amenable to standardization; they will define the place of unique individual genius and the criteria of talent that cannot be taught'*.

Larson is indebted to the work of Jamous and Peliolle (1970). Following Weber's insights into the nature of modernity, these authors recognised that the abstract, scientific nature of modern expert knowledge meant it was open to a process of rationalisation and codification into standardised rules and procedures. They argued that this was offset by the fact that uncertainty is ever present in the application of such knowledge, and they put forward the notion that occupations possess an 'indetermination' and 'technicality' ratio (an I/T ratio). Those occupations, classified as professions, possess a high level of indetermination at the basis of their expertise. Similar to Larson, they held that the outcomes of the application of expert knowledge are more dependent on the *'potentialities and talent of the practitioner than techniques and transmissible rules'* (Jamous and Peliolle 1970: 140). This leads to an emphasis in professional education and training on *'individual and social potentialities, experience, talent, intuition etc'* (Jamous and Peliolle 1970: 139). In short, it is the I in the I/T ratio that creates the basis for social prestige and distance between the expert and the client. This was illustrated in Chap. 2 in the context of the development of the modern medical profession and the principle of professional self-regulation through the birth of the clinic.

In examining the process by which occupations originally claim and subsequently maintain professional status, Larson (1977: 6) acknowledges that *'the goals and strategies pursued by a given group are not entirely clear or deliberate for all the members'*. This is an important point. Larson's reliance on historical documentary evidence means the concept of professional project does not refer to the day-to-day actions of individual rank and file members. Rather it refers to a generalised course of collective action initiated by organisational and institutional professional elites over a particular historical period. The value of Larson's analysis is that it highlights the key role in obtaining and sustaining a market monopoly played by the establishment of occupational control over the educational credentials associated with entry into a profession. This includes the important role the indeterminate aspects of a profession's expertise play in establishing this control as legitimate. As Chap. 2 discussed in some detail, the establishment of occupational control over educational credentials is a vitally important element of historical accounts of the development of the professions – particularly for understanding how professionalism operates ideologically as a methodology of occupational control.

The possession of exclusive control over the dissemination of its knowledge base to new members means a profession's elite organisations possess substantial bargaining power from which to negotiate a 'regulative bargain' with the state. As Allsop and Saks (2002: 6) state: *'access to formally accredited education and training is...a crucial portal on which exclusory closure is based that generates definitions of insiders and outsiders'*. For example, in his analysis of the legal profession, Burrage (1988: 228) states that *'In my judgment four goals have been constant and pre-eminent in the history of the legal profession...First, lawyers have sought to control admission to, and training for, legal practice. Second, they have tried to demarcate and protect jurisdiction within which they alone are entitled to practice. Third, they have tried to impose their own rules of etiquette, ethics or practice on one another. Finally, they have tried to defend and if possible enhance their status'*.

The Dominance of the Social Closure Model

The work of authors such as Burrage, Larson and Freidson calls attention to the fact that a professions' possession of a monopoly over the market for its services is not a neutral and straightforward consequence of its possession of esoteric expertise or a code of conduct which appears to regulate member's behaviour so their actions place clients' interests first and foremost. The neo-Weberian social closure model of professionalisation became increasingly popular amongst sociologists as it focused upon the role of professional self-interest, instead of alleged altruistic tendencies, in the initial formation and subsequent development of professions such as medicine. As shown in Chap. 3, a continued series of high-profile medical malpractice cases throughout the 1970s, 1980s and 1990s revealed the tensions between public and professional interest, as reflected in the early work of Freidson (1970a, b). The sociologist and GMC lay member Margaret Stacey (1992) used insights of Larson and Freidson to show that medical control of the GMC had led to a similar tension between medicine's concern with maintaining its professional privileges and the GMC's role in protecting the public interest.

The neo-Weberian perspective places a heavy emphasis upon undertaking a historical analysis of the development of modern medicine and the regulation of medical training and work. As Chaps. 2 and 3 noted, this reinforces the need to recognise how the closed shop nature of the culture of the medical club is tied up with its epistemological foundations in modern science in general and the biomedical model in particular. This has led to a particular form of reasoning, known as the clinical mentality being placed at the centre of the occupational structure and culture of the modern medical profession. Following Larson (1977: 17), it can be argued that medicine has long possessed an 'exclusive cognitive identity'. As noted earlier, the development of an exclusive member's only occupational identity, based upon the esoteric cognitive expertise shared by group members, formed a key part of the process by which the fledging medical profession initially sought to convert its increasingly scientific credentials into social and economic rewards during the nineteenth century.

Sociological accounts of medical knowledge, work and training highlight how doctors collectively and individually possess a cognitive exclusiveness towards outsiders, given the specialist nature of medical work and the lengthy period of time it takes to train a new member of the medical club. These also reveal that a key characteristic of the culture of the medical club is a mutual respect amongst group members for each other's clinical experience and expertise. This is reinforced by the hierarchical nature of the career structure of the medical profession in general and the organisation of local medical teams in particular. This leads to the general refusal, on behalf of juniors, to criticise publicly seniors in all but the most extreme cases, particularly if they want to work their way up the career ladder (Seabrook 2004). Sociological studies of medical training and work have repeatedly highlighted that the development of clinical acumen by trainees must be accomplished through the application of a characteristic mode of reasoning that is bound up with this feeling of exclusiveness, namely, the clinical mentality. This often transcends and takes precedence over the more formal scientific basis of medical expertise that is frequently presumed to lie at the basis of medical power.

In his now classic elucidation of the clinical mentality, Freidson (1970a, b) notes that an individual doctor's knowledge and expertise is personally acquired through direct first-hand experience over the course of her professional career. Freidson

(1970a, b: 170) holds that at the basis of the clinical mentality lies a *'kind of ontological and epistemological individualism'*. He argues that the nature of her work makes the medical practitioner a pragmatist. She is driven to draw upon her experience of previous similar concrete clinical cases when making her professional judgements, instead of utilising more formal resources such as clinical protocols or statistical evidence. Freidson holds that this pragmatism comes about from a doctor's need to take action and make clinical decisions in complex practice situations so she can make a positive difference (or at least do no further harm) to the lives of the patients she is professionally responsible for. Indeed, Freidson (1970a, b: 170) says that *'in having to rely so heavily upon his personal, clinical experience with concrete, individual cases...the practitioner comes essentially to rely on the authority of his own sense, independently of the general authority of tradition or science. After all, he can only act on the basis of what he himself experiences, and if his own activity seems to get results, or at least no untoward results, he is resistant to changing it on the basis of statistical or abstract consideration. He is likely to need to see or feel the case himself'*.

Freidson (1970a, b) is not alone in making the point that the expertise of the medical profession is made up of formal-determinate and tacit-indeterminate dimensions (i.e. Allsop and Mulcahy 1996; Stacey 1992, 2000) as well as holding that it is the latter, rather than the former, that is often ultimately used by doctors to justify clinical decisions. In his discussion of the management of surgical errors argues that doctors possess two distinct 'warrants for action': 'the academic' and 'the personal'. He describes how a doctor's personal 'clinical acumen' or 'clinical expertise' is often used to 'trump' academic knowledge.

As I have discussed earlier, Becker's (1961), Atkinson (1981) and Sinclair (1997) studies of medical education similarly show that during clinical training, personal experience is often rhetorically contrasted by clinical teachers with the more formal medical knowledge trainees find enshrined in course textbooks. For example, Becker (1961: 225) notes that *'even though it substitutes for scientifically verified knowledge, it [experience] can be used to legitimate a choice of procedures for a patient's treatment and can even be used to rule out use of some procedures that have been scientifically established'*. The veneration by members of the medical fraternity of the autonomy of the individual practitioner and the existence of clinical judgement and expertise accounts for the presence of variation in clinical diagnosis and treatment, as well as the fact that medical practitioners can be collectively and individually resistant to innovation and change. However, it also leads to a shared belief amongst medical club members that they can legitimately exclude outsiders from judging members of the club. For there is a mutual recognition between club members that the inherent uncertainty at the basis of their expertise means that it is a case of 'there but for the grace of god go I' when medical errors occur. They therefore collectively 'close ranks' to ensure club members are protected.

The highly personal but mutually shared nature of the clinical mentality, alongside the inherently insular nature of medicine's 'members only' regulatory club, leads to a natural reluctance on behalf of individual members to report any concerns they may have about other club member's competence. Not least of all because club members fear of being ostracised by their peers and their careers consequently blighted. Furthermore, this situation has led to tendency within medical training for teaching by humiliation, particularly when trainees make common clinical errors (Sinclair 1997). A growing body of sociological literature reported medical students' experiences of being bullied, shouted at and publicly humiliated (Silver and Glicken 1990; Schubert 1998; Seabrook 2004). This is in spite of the fact that medicine's elite institutions have recognised that they must promote an occupational culture that is more open and accountable and encourages individual practitioners to learn from their mistakes (Catto 2006, 2007). For medicine's new professionalism requires doctors to report medical errors, whether or not they are made by themselves or their peers, and actively admit to mistakes and learn from them (Irvine 1997, 2003, 2006).

Given the preceding discussion, it should not be surprising to learn then that the neo-Weberian viewpoint has dominated the sociological study of professional regulation for the last four decades. In addition to the important insights it offers into the nature of the clinical mentality and the fundamentally exclusory nature of club governance, it encapsulates the sociolegal and political realities of the regulatory context with regard to the professions in general and medicine in particular (Stacey 1992; Moran and Wood 1993, Johnson et al. 1995; Allsop and Saks 2002; Davies 2004; Allsop 2002). The interrelated concepts of professional project, occupational monopoly and social closure reflect the reality of state licensure, as achieved by professions such as medicine, in the Anglo-American context (McDonald 1995; Elston 2004). Additionally, although the exact process by which an occupation becomes a profession (i.e. professionalisation) differs between nations and occupations, the general form of state licensing of professional groups in the Anglo-American context has historically been based upon *'the model of the medical profession of the nineteenth century...In this respect, all the health professions are licensed by statute, and the terms of the license may be modified by parliament'* (Allsop and Saks 2002: 7).

Of course, an interesting question is if the neo-Weberian viewpoint will continue to encapsulate the medical regulatory context for the foreseeable future. Here, it should be noted that, if anything, the 2008 Health and Social Care Act reinforced that medical control of the GMC will remain in some form and medical elites such as the royal colleges will continue to take the lead in controlling entry onto and exit from the register of approved medical practitioners. Subsequent chapters will flesh out this argument through examining the implementation of revalidation and reforms in the hearing of fitness to practise cases. However, the neo-Weberian perspective is not beyond criticism. It can be accused of being as one sided as early functionalist accounts when they uncritically accepted the altruistic claims made by occupational groups such as medicine. For the neo-Weberian viewpoint does highlight how professions sought to obtain, protect and promote their self-interest over the interest of their clients.

Nevertheless, it can be argued that it does so by neglecting that the day-to-day activities of a large number of health-care practitioners demonstrate that they possess a strong personal commitment to their work. Indeed, they often place their personal needs second to their professional commitments in order to ensure that patients receive the best quality of care possible. It could equally be argued, however, that the value of the neo-Weberian analysis lies in the fact that it reinforces the need for the general public and state to recognise that doctors need to be able to exercise discretion in their work and indeed can by and large be trusted to place their client's interests before their own while at the same time reinforcing to doctors that the possession of a distinctive mixture of cognitive and altruistic characteristics does not in itself justify the extent to which they have traditionally been left alone to manage their own affairs. Nevertheless, it is important to also outline the significant and important contribution to the sociological study of medical regulation made by the emergence of feminist and neo-Marxist critiques of medical power. Not least of all because this will set the scene for Chap. 5 which focuses on how contemporary challenges to medical autonomy reinforce the value of adopting a more holistic approach which draws on a range of perspectives, not just the neo-Weberian social closure viewpoint, when theorising both the immediate and long-term consequences of such developments.

The Feminist Critique

By the end of the 1970s, there was a growing feminist critique of how the professions sustained gender inequalities in society. Undoubtedly, a certain gender blindness existed in the neo-Weberian view of the professions. For example, Spencer and Podmore (1986) argued that sociological accounts of the legal profession ignored that female solicitors were marginalised by their male colleagues and discussed how this was related to broader social expectations regarding appropriate male and female roles and relationships. Their empirical research found that discrimination against women within the legal profession occurred primarily because the confrontational nature of court hearings meant law was held to be a masculine, aggressive occupation. Female solicitors were defined by their male colleagues as 'the other' through engaging in gender-laden discourses that variously categorised them as sex objects, different and unfeminine, overemotional or not tough enough. This situation enabled the allocation of female solicitors into what were seen as gender-appropriate careers, such as family law, and actively excluded them from elite occupational positions within the profession. At the time of Spencer and Podmore's study in the mid-1980s, only 2% of judges were women while they were no women law lords. This was preventing them from becoming a part of the legal professions self-regulatory elite (Dingwall and Lewis 1983).

As noted earlier, a similar situation was found by Stacey (1992) in her study of the GMC. The first female member of the GMC was not elected until the 1950s, that is, nearly 100 years after the foundation of the GMC in 1858. There were only three female members of the GMC throughout the 1970s and early 1980s (two of whom were non-medical, including Stacey herself). While there was an over-representation of female doctors in what were seen within the profession as being female friendly specialties, such as general practice. As in the case of the legal profession, female friendly specialties were not conducive to obtaining access to the higher echelons of the professions elite training and regulatory institutions.

Feminism is a not a unitary social theory. It incorporates authors operating from liberal, radical, neo-Marxist, Black and postmodern viewpoints, to name a few (Anthias and Yuval-Davis 1993). The concept of patriarchy has traditionally been at the centre of feminist viewpoints, with its claim that there is an all-pervasive male gaze, which directly oppresses women and possesses institutionalised power within the apparatus of the state. This has been criticised by feminists and postmodern thinkers who hold an anti-essentialist view of the self and so reject the idea that there is a universal female subject or a common feminine experience and identity (Barrett and Philips 1992). Authors working in the field of men's studies extend these views further and use the notion of hegemonic masculinity to explore the oppressive features of the rational, domineering, aggressive and exploitative White Anglo-Saxon Protestant male (Connell 1995). Throughout the 1970s and 1980s, a growing number of authors interested in the social role of medicine and working from a feminist perspective focused upon the fact that the history of modern medicine and its treatment of women was tied up with a broader narrative of subordination of 'the female' to 'the male'. Women were socially constructed as 'the other' and assigned normative social roles belonging to the private sphere. For example, studies by Barker-Penfield (1979) and Holmes (1980) traced the historical development of the medical discipline of obstetrics and gynaecology. They highlighted how this was tied up with a socio-economic need to manage the female body to locate its biological destiny within the social roles of mother and housewife.

The work of Pfeffer (1985) shows that the form of language used in medical textbooks to describe common 'female conditions' constructs the female body as being a poor second to its healthier male counterpart. Pfeffer's work discusses how a women's experience of her body is mediated through medical categories and conditions that possess fundamentally negative gender images, as 'infantile' uterus, 'failed' labour, placental 'insufficiency', 'irregular' menstrual cycles and hormonal 'imbalances'. Underlying the critique of the feminist perspective was the belief that the structure of medical knowledge was in many ways sexist and patriarchal due to medicine's close relationship to science (Fox-Keller 1985). The feminist perspective holds that the rationality of 'the enlightenment', which spawned modern scientific thought, was inherently masculine. Women were perceived as 'the other' and held to be illogical or irrational. Women were fundamentally flawed and emotional creatures inextricably bound to their reproductive role (Ehrenreich and English 1973). As Fox-Keller (1985: 78) maintains, the feminist perspective held that *'in characterizing scientific and objective thought as masculine, the very activity by which the knower can acquire knowledge is gendered'*.

Ehrenreich and English (1979) documented how women were socially defined by society as fragile creatures that were prone to hysteria. The development of scientific medicine over the course of the nineteenth century allowed the source of this problem to be increasingly located within the female reproductive system. This explanation was seen as socially acceptable as it precluded men from the possibility of becoming hysterics. To this day, female patients more than male patients are likely to be viewed as unhappy, depressive and anxious by general practitioners and psychiatrists. Furthermore, they are more likely to receive pharmacological treatments such as Valium and Prozac and Serotoxin (Prior 1999).

The feminist critique of medical knowledge and practice extended into the very organisation of the medical profession. Witz (1992) argued in line with the neo-Weberian thesis that the medical profession obtained its market monopoly in the United Kingdom through using its educational credentials as a 'bargaining chip' from which to negotiate with state control over its regulatory arrangements. She also held that medicine's achievement of 'social closure' in the nineteenth century succeeded because the strategy of closing off medical training and practice to all but an elite few was in line with boarder social norms of the time. Medicine actively sought to keep women in the private not public sphere. It excluded women and to a lesser extent working-class men from practicing medicine due to its historically close association with the gentry. She argues that medicine's elite institutions used, firstly, exclusionary strategies to deny women entry into medical school and so the medical register, and secondly, demarcatory strategies where medical control was firmly established over other health-care occupations dominated by women, such as nursing and midwifery.

As discussed earlier, medicine's professionalisation process was certainly dominated by men and involved the appropriation of healing and caring domains that had traditionally belonged to women. Far from being a neutral science, medicine reflected the patriarchal and class-based nature of society at the time. By the mid-1980s, the work of feminist authors such as Ann Oakley (1984) had made a significant contribution to the growing recognition that the practice of modern medicine was largely socially and culturally bounded. Authors working within neo-Weberian and feminist positions shared a common view of medical practice, which was diametrically opposed to medicine's self-image as scientific, value-free and morally neutral (Elston 2004).

However, despite the important contribution of the feminist perspective, the neo-Weberian continued to dominate sociological analysis of professionalism and the principle of professional self-regulation in the Anglo-American literature. The feminist perspective tended to be held by sociologists concerned with the sociological analysis of the professions to supplement and expand the neo-Weberian perspective, not necessarily replace it (Lupton 1994). The was because there was an ongoing debate within sociology about the extent to which the clinical gaze of modern medicine was a social construct and therefore could be seen as inherently gendered. Some commentators held the view that although a human undertaking, and therefore open to a range of intervening socio-economic and cultural factors, medical science does reflect a reality that exists 'out there' independent of the observer. Modern medical expertise and technology consequently was seen to exist beyond the particular circumstances surrounding their creation and application. Elston (1991) notes that medical judgements are likely to be seen as valid and true because modern medicine possesses a considerable amount of cultural authority over definitions of reality. This is due to the predictive power of the randomised clinical trial. However, other social commentators held that it is impossible to trust the objectivity and neutrality of the seeing-knowing subject whose gaze extracts knowledge from the world, whether they are a scientist, a doctor or a philosopher.

The growing social constructionist influence of postmodernism gave weight to the view that human knowledge of the world is limited by language. Indeed, this asserts that it is impossible to apprehend reality outside of the arbitrary linguistic conventions and metaphoric imperatives belonging to the language games used to describe it (Drolet 2004). Most sociologists, like Turner (1995), held onto the middle ground within this debate. He argued that some aspects of modern medical knowledge and expertise, for example, the diagnosis and treatment of hyperactivity in children, are more clearly socially constructed than others, for example, cirrhosis of the liver, as these possess a structural and biochemical origin within the human body.

Adopting a somewhat pragmatic position in this manner has meant that to this day, it is possible to detect, as Riska (2001) discusses, three possible stances towards the question of whether or not medical knowledge and work are too gendered. First is the view that medicine is a gender-neutral activity, and the creation and application of medical knowledge and expertise is a value-free affair. As Riska (2001) notes, this viewpoint is not so much held by feminist authors themselves but by members of the public as well as many doctors themselves. Second is the view that medicine is an inherently masculine activity, which promotes a negative view of women and the female body. It relegates women to the private sphere and the role of mother and wife due to their reproductive role and biological difference to the 'alpha male'.

Riska (2001) notes that the third position operates somewhere in the middle of these two extremes. This view holds that the creation and application of modern scientific medical knowledge does appear to mirror the nature of the world in which human beings live. Furthermore, it does enable them to access and actively manipulate the biological realties of their existence. This position also asserts the practice of medicine is nevertheless an inherently social activity; it reflects the broader cultural values of the society within which it operates. Therefore, medicine possesses gendered processes and practices. Riska (2001) provides cross-national evidence to show that although more women today than ever before are pursing medicine as a career, a glass ceiling still operates inside the medical club that stops female doctors accessing certain prestigious surgical sub-fields. Nevertheless, they are two key interrelated reasons why sociological analysis of professionalism has continued to be dominated by the neo-Weberian viewpoint. First, the neo-Weberian viewpoint by and large reflects that nature of the regulatory context in the United Kingdom in regard to health and social care professions in general and the medical profession in particular (Stacey 1992; Moran and Wood 1993; Johnson et al. 1995; Allsop and Saks 2002; Davies 2004; Allsop 2002; Slater 2007). Indeed, medical control over admission onto and exit from a state-approved register of practitioners will continue for the foreseeable future. It is precisely because of this fact that feminist authors working within the UK context, such as Elston (1991, 1997, 2004), Stacey (1992, 2000) and Witz (1992), used the framework provided by neo-Weberian viewpoint when analysing how medicine as a profession is regulated. The second reason why the feminist position did not replace the neo-Weberian viewpoint was that its critique of medicine remained focused upon forms of medical knowledge and technology that were experienced mainly by women, such as reproductive technology (Lupton 1994; Elston 1997). Although this focus was justifiable, it nevertheless limited the ability of the feminist perspective to contribute to broader sociological debates regarding the regulatory arrangements concerning occupational groups categorised as professions. Because its research focus was 'gender exclusive', the contribution of the feminist perspective lies within the broader field of the sociology of health and illness and not within the sociological study analyse of professional self-regulation (Nettleton 1995). Particularly as this is restricted to the analysis of occupational groups which claim to possess not just esoteric specialist knowledge but also an ethical code of conduct that requires, they place their clients' interests before their own (McDonald 1995). This code of conduct is used to obtain not only social and economic rewards but also exclusive occupational control over members training, practice and discipline (Freidson 1970a, b, 1994, 2001). This includes traditionally female-dominated occupations such as nursing (Stacey 1988).

The history of the professionalisation of nursing in the UK shows how broader social norms alongside the restrictive actions of a male-dominated medical profession initially blocked nurse's claim to professional status. That is, until a mixture of NHS service needs and continued political activism on behalf of nurses throughout the 1950s and 1960s eventually lead the establishment of a General Nursing Council in 1979 (Riska and Wegar 1993). Yet historical narrative shows that throughout this time, nursing sought proactively to exclude third-party evaluation of practitioners' activities as a key part of its quest for professional status. Indeed, like the medical profession before it, nursing eventually acquired a legal statute through parliament that enshrined in law its right to possess exclusive occupational control over a register of member's entry into and exit from the nursing profession, as well as the standards governing members' training, practice and discipline (Davies and Beach 2000). Furthermore, similar to other occupations categorised as professions (e.g. medicine, law, psychiatry and social work), nursing's self-regulatory body has in the last two decades been accused of being elitist, inherently inward looking and protectionist, as a result of high-profile malpractice cases in the media. This has led to calls for greater lay involvement in nurse regulation and a more open and multidisciplinary approach towards nurse training and discipline.

The Neo-Marxist Critique

Despite its dominance in the sociological study of professional regulation, the neo-Weberian perspective was criticised by authors operating from a neo-Marxist viewpoint for failing to account for the entwined nature of the development of the modern state and professions such as medicine, as was touched upon earlier when discussing club governance (Moran 1999, 2004). Indeed, although his work was (and still is) regarded as a sociological classic, Freidson was criticised by neo-Marxist commentators for ignoring the political economy and under theorising the relationship between medical and state power. As Larson (1977: xiv) notes, Freidson's work does tend to assume that the professions are 'independent from or at least neutral vis-à-vis the class structure'. In contrast, the neo-Marxist perspective of the professions argued that medical dominance in the health-care division of labour played a central role in the surveillance and reproduction of working-class labour on behalf of capital (Johnson 1977). As Johnson (1977: 106) notes: *'the professionalism of medicine – those institutions sustaining its autonomy – is directly related to its monopolization of 'official' definitions of illness and health. The doctor's certificate defines and legitimates the withdrawal of labor. Credentialism, involving monopolistic practices and occupational closure, fulfills ideological functions in relation to capital and reflects the extent to which medicine in its role of surveillance and the reproduction of labor power is able to draw upon powerful ideological symbols'*. Here McKinley is typical of the neo-Marxist viewpoint when he states, *'the House of Medicine under capitalism will never contribute to improvements in health unless such improvements facilitate an acceptable level of profit'* (McKinley 1977: 462).

According to neo-Marxists, there is no difference between the production of taken-for-granted capitalist commodities such as cars, fridges and clothes and the practice of the surgical techniques of modern medicine, such as open-heart surgery (Navarro 1976). Both involve the search for profit. Large corporations involved in the production of medical supplies, particularly pharmaceutical therapies, profit from individual experiences of illness and disease (Navarro 1986). Neo-Marxist commentators may agree with their neo-Weberian counterparts that medicine possessed substantial control over other health-care occupations and patients. Nevertheless, they also held that medical work was increasingly coming under direct bureaucratic-managerial surveillance and control operating on behalf of capital (McKinley 1977) – a point that will be returned to in Chap. 5 when the proletarianisation thesis is discussed.

The neo-Marxist sociologist Navarro (1976, 1986) argued that medical autonomy is tied to the needs of capital. He held that it only emerged because the increasingly scientific foundations to medical expertise were congruent with the interests and needs of nineteenth century industrialists, who were using the apparently neutral concept of science to justify the introduction of new factory-based mass production methods. Navarro (1976: 31) argued that there had been an *'invasion of the house of medicine by capital'*, and consequently, medical knowledge and technology could not be seen as separate from capitalism but rather was part of it. Medical knowledge was not overlain onto capital ideology, but rather modern medicine under capitalism is capitalist medicine (Navarro 1980). Navarro views medicine's essentially mechanistic view of the human body as being tied up with the capitalist mode of production. Neo-Marxists argue that medicine plays a key role in supporting the status quo in the capitalist system by reinforcing the idea that lifestyle choices as well as natural processes are responsible for personal and collective experiences of illness and disease. They hold that in adopting this approach, medicine camouflages alternative social and economic factors relating to worker exploitation under the capitalist system (McKinley 1977). They follow Marx's colleague, Fredrick Engels, who in his key text *The Condition of the Working Class in England* (1974) held that an individual's personal experience of, for example, alcoholism was an outcome of the impoverished life chances available to low-paid workers living in the slums of industrialised cities. For Engels, dependence on alcohol was a result of an attempt to 'blot out' the harsh reality of the working and living conditions present in nineteenth century society. It was not due to some inherent biological tendency towards addiction. Waitzkin's (1989) work on how doctor-patient interaction reinforces class inequalities focuses upon this point. Waitzkin (1989: 223) argues that during the doctor-patient encounter, *'technical statements help direct patients' responses to objectified symptoms, signs and treatment. This reification shifts attention away from the totality of social relations and the social issues that are often causes of personal troubles'*.

A key criticism of the neo-Marxist viewpoint is that similar to functionalism it seeks to explain medicine's position in society as stemming from the important social role it plays in maintaining the established social order. The main difference between the two perspectives is that neo-Marxists regarded this order as exploitative and ultimately offering no benefit to the individual worker. This is an overly simplistic viewpoint. In contrast, authors operating from the Foucauldian governmentality perspective may like their neo-Marxist counterparts focus upon how health and social care professions such as medicine are deeply bound up with the process of governing populations. So much so that governmentality authors such as Johnson (1995: 13) hold that, *'the expert is not sheltered by the environing state, but shares in the autonomy of the state'*.

Yet the key difference between the respective neo-Marxist and governmentality perspectives is that while the neo-Marxist viewpoint sees this state of affairs as fundamentally repressive, by arguing it sustains class-based inequalities, in contrast a governmentality viewpoint considers its productive affects. It does this by focusing upon the role professional expertise plays in promoting and sustaining an individuals' capacity for engaging in self-surveillance and self-regulation (i.e. through acting on advice provided by their local general practitioner and other public health experts regarding appropriate dietary and exercise regimes). For the governmentality perspective sees this as being part of the ability of expertise to render *'the complexities of modern social and economic life knowable, practicable and amenable to governing'* (Johnson 1995: 23). The governmentality perspective and its contribution to the sociological study of the professions and professional regulation is discussed in more detail in Chap. 5 as this will lead us into the analysis of how contemporary developments within the regulation of medicine can be theorised through the lens of the restratification thesis. As we shall see, this will set the scene for the analysis of revalidation in Chap. 6 and the hearing of fitness to practise cases in Chap. 7.

Conclusion

This chapter has outlined the development of the sociological analysis of the professions and medical regulation. It noted how although early sociological analysis of the professions focused on the esoteric nature of expertise possessed by certain occupations within society, it was particularly concerned with their claim to place the interests of their clients before their own self-interest. This, it was felt, played an important role in maintaining social order and restraining self-interest within capitalistic social systems. This chapter then discussed how from the 1970s onwards first neo-Weberian and later feminist and Marxist authors were heavily critical of this earlier viewpoint and highlighted how professions such as medicine in fact used their knowledge and expertise to obtain a market monopoly, exclude competitors and obtain economic and social status reward. Furthermore, it was noted that far from being an altruistic discipline, medicine in fact plays a key role in the governance of population and masking the negative effect of capitalistic modes of production on the health and welfare of the population. This close relationship between medicine and the state will be explored in greater detail in Chap. 5 which examines the governmentality viewpoint. The end-of-chapter self-study activities will help you to consolidate what you have learnt before moving on to consider the contents of Chap. 5.

Self-Study Activity

1. Write a 1,000-word essay which critically evaluates the strengths and weaknesses of the functionalist sociology analysis of occupational groups classified as professions and its focus on the altruistic role they play in maintaining social order within society.
2. Produce a 15-min PowerPoint presentation which critically considers the neo-Weberian perspective in relation to how professional groups such as medicine use their expertise and self-proclaimed altruistic tendencies to obtain a market monopoly for their services.
3. Write a 100-word essay which critically evaluates the contribution of the feminist perspective to the sociological study of medical regulation in general and the neo-Weberian critique of medical power in particular.

References

Abbott, A. (1988). *System of professions: An essay on the expert division of labor*. Chicago: Chicago University Press.

Allsop, J. (2002). Regulation and the medical profession. In J. Allsop & M. Saks (Eds.), *Regulating the health professions* (pp. 111–133). London: Sage Publications.

Allsop, J., & Mulcahy, L. (1996). *Regulating medical work: Formal and informal controls*. Milton Keynes: Open University Press.

Allsop, J., & Saks, M. (Eds.). (2002). *Regulating the health professions*. London: Sage Publications.

Anthias, F., & Yuval-Davis, N. (1993). Contextualizing feminism: Gender. *Ethnic and Class Divisions Feminist Review, 15*, 62–75.

Atkinson, P. (1981). *The clinical experience: The construction and reconstruction of medical reality*. Farnborough: Gower.

Barber, B. (1963). Some problems in the sociology of the professions. *Daedalus, 92*, 669–689.
Barker-Penfield, J. (1979). Sexual surgery in the late nineteenth century. In C. Dreifus (Ed.), *Seizing our bodies*. New York: Vintage.
Barrett, M., & Philips, A. (1992). *Destabilizing theory*. Cambridge: Polity Press.
Becker, H. S. (1961). *Boys in white: Student culture in medical school*. Chicago/London: University of Chicago Press.
Becker, H. S., Geer, B., Hughes, E. C., & Strauss, A. L. (1961). *Boys in white: Student culture in medical school*. Chicago: University of Chicago Press.
Berlant, J. L. (1975). *Profession and monopoly: A study of medicine in the United States and Great Britain*. Berkeley: University of California Press.
Burrage, M. (1988). Revolution and the collective action of the French, American and English legal professions. *Law and Social Enquiry: The Journal of the American Bar Foundation, 13*(2), 225–277.
Carr-Saunders, A. M., & Wilson, P. A. (1933). *The professions*. Oxford: The Clarendon Press.
Catto, G. (2006). *GMC news (3)*. London: GMC.
Catto, G. (2007). Will we be getting good doctors and safer patients? *British Medical Journal, 334*, 450.
Connell, R. W. (1995). *Masculinities*. London: Allen and Unwin.
Davies, C. (2004). Regulating the health care workforce: Next steps for research. *Journal of Health Services Research & Policy, 9*(1), 2–10.
Davies, C., & Beach, A. (2000). *Interpreting professional self-regulation: A history of the United Kingdom central council for nursing midwifery and health visiting*. London: Routledge.
Dingwall, R., & Lewis, P. (Eds.). (1983). *The sociology of the professions: Lawyers, doctors and others*. London: McMillan.
Drolet, M. (Ed.). (2004). *The postmodernism reader*. London: Routledge.
Durkheim, E. (1957). *Professional ethics and civic morals*. London: Routledge and Kegan Paul.
Ehrenreich, B., & English, D. (1973). *Witches, midwives and nurses: A history of women healers*. London: Writers and Readers Publishing Cooperative.
Elston, M. A. (1991). The politics of professional power: Medicine in a changing medical service. In J. Gabe, M. Calman, & M. Bury (Eds.), *The sociology of the health service* (pp. 187–213). London: Routledge.
Elston, M. A. (1997). *The sociology of medical science and technology*. Oxford: Blackwell.
Elston, M. A. (2004). Medical autonomy and medical dominance. In J. Gabe, M. Bury, & M. A. Elston (Eds.), *Key concepts in medical sociology* (pp. 167–182). London: Sage Publications.
Engels, F. (1974). *The condition of the working class in England*. London: Progress Publishers.
Etzioni, A. (1969). *The semi-professions and their organization*. New York: Free Press.
Fox-Keller, E. (1985). *Reflections on gender and science*. New Haven: Yale University Press.
Freidson, E. (1970a). *The profession of medicine*. New York: Dodds Mead.
Freidson, E. (1970b). *Professional dominance: The social structure of medical care*. New York: Atherton.
Freidson, E. (1994). *Professionalism reborn: Theory, prophecy and policy*. Cambridge: Polity Press.
Freidson, E. (2001). *Professionalism: The third logic*. Cambridge: Polity Press.
Gerth, H. H., & Wright Mills, C. (1946). Introduction. In M. Weber (Ed.), *Essays in sociology*. Oxford: Oxford University Press.
Holmes, H. (1980). *Birth control and controlling birth*. Clifton: Humana Press.
Hughes, E. (1963). Professions. *Daedalus, 92*, 655–668.
Irvine, D. (1997). The performance of doctors 1: Professionalism and self-regulation in a changing world. *British Medical Journal, 314*, 1540–1550.
Irvine, D. (2003). *The doctors tale: Professionalism and the public trust*. London: Radcliffe Medical Press.
Irvine, D. (2006). Success depends upon winning hearts and minds. *British Medical Journal, 333*, 965–966.

Jamous, H., & Peliolle, B. (1970). Changes in the French university-hospital system. In J. A. Jackson (Ed.), *Professions and professionalization* (pp. 144–165). Cambridge: Cambridge University Press.
Jewson, N. D. (1974). Medical knowledge and the patronage system in eighteenth century England. *Sociology, 8*, 369–385.
Jewson, N. D. (1976). The disappearance of the sick-man from the medical cosmology. *Sociology, 10*, 225–244.
Johnson, T. J. (1972). *Professions and power*. London: MacMillian.
Johnson, T. J. (1977). Professions in the Cass structure. In R. Scase (Ed.), *Class cleavage and control* (pp. 156–177). London: Allen and Unwin.
Johnson, T. J. (1995). Governmentality and the institutionalization of expertise. In T. Johnson, G. Larkin, & M. Saks (Eds.), *Health professions and the state in Europe* (pp. 177–193). London: Routledge.
Johnson, T., Larkin, G., & Saks, M. (Eds.). (1995). *Health professions and the state in Europe*. London: Routledge.
Larkin, G. (1983). *Occupational monopoly and modern medicine*. London: Tavistock.
Larson, M. S. (1977). *The rise of professionalism: A sociological analysis*. Berkeley: University of California Press.
Lupton, D. (1994). *Medicine as culture: Illness, disease and the body in western societies*. London: Sage Publications.
McDonald, K. M. (1995). *The sociology of the professions*. London: Sage Publications.
McKinlay, J. B. (1977). The business of good doctoring or doctoring as good business: Reflections on Freidson's views on the medical game. *International Journal of Health Services, 7*, 459–483.
Moran, M. (1999). *Governing the health care state; a comparative study of the United Kingdom, the United States and Germany*. Manchester: Manchester University Press.
Moran, M. (2004). Governing doctors in the British regulatory state. In A. Gray & S. Harrison (Eds.), *Governing medicine: Theory and practice* (pp. 133–155). Milton Keynes: Open University Press.
Moran, M., & Wood, B. (1993). *States, regulation and the medical profession*. Milton Keynes: Open University Press.
Murphy, R. (1988). *Social closure*. Oxford: Clarendon.
Navarro, V. (1976). *Medicine under capitalism*. New York: Prodist.
Navarro, V. (1980). Work ideology and medicine. *International Journal of Health Services, 10*, 523–550.
Navarro, V. (1986). *Crisis, medicine and health: A social critique*. New York: Tavistock.
Nettleton, S. (1995). *The sociology of health and illness*. Cambridge: Polity Press.
Oakley, A. (1984). *The captured womb*. Oxford: Blackwell.
Parsons, T. (1949). *Essays in sociological theory*. Glencoe: Free Press.
Pfeffer, N. (1985). The hidden pathology of the male reproductive system. In H. Homans (Ed.), *The sexual politics of reproduction* (pp. 93–113). Aldershot: Gower.
Prior, P. (1999). *Gender and mental health*. London: Macmillan.
Riska, E. (2001). *Medical careers and feminist agendas: American, Scandinavian and Russian women physicians*. London: Sage Publications.
Riska, E., & Wegar, K. (1993). *Gender, work and medicine women and the medical division of labor*. London: Sage Publications.
Schubert, M. K. (1998). The relationship between verbal abuse of medical students and their confidence in their clinical abilities. *Academic Medicine, 73*(8), 907–909.
Seabrook, M. (2004). Intimidation in medical education: Students and teachers perspectives. *Studies in Higher Education, 29*(1), 59–74.
Silver, H. K., & Glicken, A. D. (1990). Medical student abuse: Incidence, severity and significance. *Journal of the American Medical Association, 263*, 527–532.
Sinclair, S. (1997). *Making doctors: An institutional apprenticeship*. London: Berg.
Spencer, A., & Podmore, D. (1986). *In a man's world*. London: Tavistock.

Stacey, M. (1988). *The sociology of health and illness*. London: Uwin Hymen.
Stacey, M. (1992). *Regulating British medicine*. London: Wiley.
Stacey, M. (2000). The general medical council and professional self-regulation. In D. Gladstone (Ed.), *Regulating doctors* (pp. 15–31). London: Institute for the Study of Civil Society.
Tawney, R. H. (1921). *The acquisitive society*. New York: Harcourt Brace.
Turner, B. S. (1995). *Medical power and social knowledge* (2nd ed.). London: Sage Publications.
Weber, M. (1947). On Sociology, in H. H. Gerth and C. Wright Mills (Eds.), *Essays in Sociology*. Oxford: Oxford University Press.
Weber, M. (1978). *Economy and society*. London: The University of California Press.
Waitzkin, H. (1989). A critical theory of medical discourse. *Journal of Health and Social Behaviour, 30*, 220–239.
Witz, A. (1992). *Professions and patriarchy*. London: Routledge.

Chapter 5
Sociological Deconstructions II: Governmentality and Restratification

Abstract This chapter continues to focus on the development of the sociological study of the professions and professional regulation as initially outlined in Chap. 4. Leading on from the critique of the neo-Marxist perspective outlined in Chap. 4, this chapter begins by discussing the contribution of Foucault and the governmentality perspective to the study of the professions and professional regulation. Here, the chapter highlights how the governmentality perspective reinforces the need to place analysis of reforms in medical regulation, as well as professional regulation more generally, against the broader background of shifts in how good government is conceptualised and practised within neo-liberal western nation-states. In doing so, this chapter notes how the emergence of calls for more transparent and accountable performance management systems in relation to professional regulation is bound up with the emergence of the risk saturated conditions associated with the risk society. Consequently, rather than seeing a whole-scale decline in medical autonomy, it is argued we are instead seeing a slight compression of its borders alongside a shift in the conditions under which it can be practised, with the result that its enactment becomes a more contested exercise in performance management, particularly when medical error occurs. Here, the chapter notes the similarities between the governmentality and neo-Weberian perspectives via analysis of the literature surrounding the question of if medical autonomy is in decline. In doing so, this chapter introduces the restratification thesis and discusses how rank-and-file practitioners appear to be coming under greater surveillance and control from elite peers as they seek to maintain the principle of professional self-regulation in the face of calls from patient and managers for increased medical transparency and accountability. This sets the scene for exploration of the impact of the restratification thesis on professional practice via analysis of revalidation in Chap. 6 and the handling of fitness to practise cases in Chap. 7. End-of-chapter self-study tasks are provided so the reader can engage in further study in relation to chapter contents.

J.M. Chamberlain, *The Sociology of Medical Regulation: An Introduction*,
DOI 10.1007/978-94-007-4896-5_5,

Introduction

Chapter 4 ended by noting the continued dominance of the neo-Weberian perspective and discussing the continued relevancy of the neo-Marxist position for the study of professional regulation. For neo-Marxist commentators such as McKinley (1977) and Navarro (1980), medicine and the state act in unison to govern society in the interests of capital. Whatever the strengths and weaknesses of such a position, as was first discussed in Chap. 3 when the relationship between the medical club and the broader club governance model in operation during the nineteenth century was outlined, it does seem to be the case that somehow doctors share in the autonomy of the state. Insofar that the nature of medical autonomy reflects the prevailing governing conditions, modern states find themselves in.

Against this background, this chapter outlines and discusses the work of Foucault, paying particular attention as it does so to the concept of governmentality and the promotion of the broader social conditions associated with the re-emergence of the enterprise self as a governing tool within modern western liberal democratic states. Here, the importance issue of the possible decline of medical autonomy will be explored in relation to the respective proletarianisation and deprofessionalisation theses. These theses were proposed as it increasingly became apparent from late 1970s onwards that medicine as a profession was being challenged by patients and managers while at the same time becoming subject to greater state scrutiny and performance management via legislative frameworks introduced under the banner of health service reform to increase consumer choice and reduce medical risk. This leads the chapter to discuss the emergence of the restratification thesis and set the scene for subsequent discussion in later chapters of revalidation and the handling of fitness to practise cases. But before any of this can be done, it is first necessary to discuss the governmentality perspective and its contribution to the sociological study of the professions and professional regulation.

Governmentality and the Revival of Liberalism

As was noted in Chap. 2, the 1970s and 1980s saw the renewal of liberalism as an economic and political ideology, with its emphasis on individualism, advocacy of rolling back the state and belief in the ability of the discipline of the market to promote consumer choice, improve service quality and minimise risk. Classical liberalism had emerged in the seventeenth and eighteenth centuries, through the works of a variety of writers, such as Thomas Hobbes, John Stuart Mills, Adam Smith, Thomas Locke, Jeremy Bentham and Herbert Spencer. It is possible to identify at the centre of classical liberalism the underlying concept of possessive individualism (Macpherson 1962). Macpherson (1962) argues that for these thinkers, the individual and her capabilities prefigure the circumstance into which she is born. In short, her talents and who she is owes nothing to society; rather she owns herself, and she is morally and legally responsible for herself and herself alone. She is naturally self-reliant and free from dependence on others. She need only enter into relationships with others because they help her pursue her self-interests. According to this viewpoint, society is seen as a series of market-based relations made between self-interested subjects who are actively pursuing their own interests. Only by recognising and supporting this position politically and economically will the greatest happiness for the greatest number be achieved. It is for such reasons that classical liberalism is commonly held to be a critique of state reason that seeks to set limits on state power.

It is against this background of the re-emergence of liberalism that sociologists concerned with the governance of expert forms of knowledge have recognised the importance of the work of Foucault and his concept of governmentality in the analysis of the relationship between the professions and the state (Peterson and Bunton 1997). The work of Foucault (1965, 1970, 1972, 1977, 1979, 1982, 1985a, b, 1986, 1989, 1991a, b) highlights how individual subjectivities are neither fixed nor stable, but rather are constituted in and through a spiral of power-knowledge discourses. These are generated by political objectives, institutional regimes and expert disciplines, whose primary aim is to produce governable individuals (Deleuze 1988; Peters 2001). At the end of the eighteenth century onwards, there was a steady growth in 'the dubious sciences', what Foucault calls the human sciences, particularly new scientific disciplines, such as psychiatry, criminology, public health medicine and sociology. Foucault holds that a key outcome of the rise of these new sciences was the more intensive use of dividing practices to objectify an individual and their body via systems of notation, classification and standardisation. Foucault argues that through their examination and assessment techniques, experts produce normative classifications for subjective positions (normal, mad, sexually deviant, etc.) which increasingly became inscribed within the disciplinary regimes of society's organisational and institutional structures. There regimes spread throughout society as a whole as the dominance of the pastoral power of Christianity started to decline and a more secular concern with what can be termed 'the conduct of conduct', that is, governmentality, emerged from the sixteenth century onwards (Foucault 1991).

Foucault first published his study of Governmentality in 1979 (Foucault 1991b) and further developed it as a concept within a series of lectures given at the *College de France* (Burchell et al. 1991). Foucault discusses that from around the sixteenth century onwards, an ever-growing number of treatises were published on the governance of the soul and the self, the family and the state. These were published against an increasingly complex background of technological development, rapid social change and political and intellectual upheaval. It should not be surprising to learn that events such as the enlightenment, the reformation, the rise of modern science and the development of industrial capitalism collectively led to a growth in the writing of treatises which sought to answer fundamental problems of rule: *'how strictly, by whom, to what end, by what methods etc'* (Foucault 1991b: 88). Foucault notes that these treatises focused more and more upon the idea that good governance entailed the right disposition of things and had as its aim the common welfare and salvation of all. Governance came to involve securing the security, health, welfare and happiness of the population. The *'population comes to appear above all else as the ultimate end of government'* (Foucault 1991b: 100). Over time, governance would become increasingly tied into a liberalist conception of economics. Good government was economical, both fiscally and in its use of power. Furthermore, the development of new forms of expertise, Foucault's dubious human sciences, such as psychology, medicine and sociology, are tied up with this need to govern the population to ensure its betterment. This was because at an increasingly complex administrative and bureaucratic level, the population was seen as possessing its *'own regularities, its own rate of deaths and diseases, its cycles of scarcity, etc'* (Foucault 1991b: 90). Consequently, *'novel forms of expertise in the fields of public health hygiene, mental health and mass surveillance emerged in concert with developing government policies and programmes...and were intimately involved in the construction of governable realms of social reality'* (Johnson 1994: 142). The modern professions and their associated training and regulatory arrangements are emergent as an aspect of the formation of a liberal form of governmentality that has as its target the population and its welfare and which itself was emergent with the growth of capitalist industrial economies across Europe during the nineteenth century.

Foucault notes that two other forms of power, sovereignty and discipline, are tied up with the development of the power of a population-focused form of governance, with its concern for the conduct of conduct, to enable the promotion of the security, health, wealth and happiness of individual subject-citizens. Sovereign command power is exercised over subjects through the juridical and executive arms of government. Historically, sovereign power is related to monarchical rule, with its executive mechanisms of constitutions, laws and parliaments. Over time, these were made into more representative institutions through the development of democratic ideals, with allegiance to the monarch becoming transformed into allegiance to the rule of law (Foucault 1991b). The power of discipline goes back to ancient religious, military and educational practices. As Foucault noted in *Discipline and Punish* (1977), its expansion over the population during the seventeenth and eighteenth centuries is tied up with a growing administrative and institutional need to survey and make docile individual and collective bodies. Disciplined individuals have acquired habits of action and thought which enable them to act in appropriate and expected ways and to do so through the exercise of self-control (Foucault 1977). Good governance is about how to best align the sovereign power of command and productive disciplinary power in order to achieve the primary object of securing the health, wealth and happiness of the population. This is why Foucault argues that the power of governance does not replace the power of discipline or sovereignty. Rather it recruits them. Indeed, Foucault (1991b: 102) argues that *'we need to see things not in terms of the replacement of a society of sovereignty by a disciplinary society and the subsequent replacement of a disciplinary society by a society of government; in reality one has a triangle, sovereignty-discipline-government, which has as its primary target the population'*. In short, the power of governance is where *'technologies of domination of individuals over one another have recourse to processes by which the individual acts upon himself and , conversely...where techniques of the self are integrated into structures of coercion'* (Foucault 1980: 2). Governance *'retains and utilizes the techniques, rationalities and institutions characteristic of both sovereignty and discipline... [but it also]...departs from them and seeks to reinscribe them. The object of sovereign power is the exercise of authority over the subjects of the state within a defined territory, e.g. the deductive practices of levying taxes, of meting out punishments. The objects of disciplinary power is the regulation and ordering of the numbers of people within that territory e.g. in practices of schooling, military training or the organization of work. The new object of government, by contrast, regards these subjects and the forces and capacities of living individuals, as members of a population, as resources to be fostered, to be used and to be optimized'* (Dean 1999: 12).

The Natural Order of Liberal Governance

The institutionalisation of professionalism as a self-regulatory strategy was ensured by liberalism's focus upon what is natural and what is not. The development of modern, rational and scientific expertise is entwined with the growth of the view that personal freedom is the natural state of humankind, and minimal forms of government are the natural way of things (Rose 1999). Governmentality seeks the optimum method by which to affect at a distance the way individuals conduct themselves without recourse to direct forms of repression or intervention, that is, unless they are absolutely necessary (Barry et al. 1996). This is because the effectiveness of liberal mentalities of rule lies in their ability to align, *'the objectives of authorities wishing to govern and the personal projects of those organizations, groups and individuals who are the subjects of government'* (Rose 1999: 48). A key method by which this goal was achieved from the nineteenth century onwards (and still today) was by harnessing the expertise of doctors, lawyers, teachers, etc. *"into the process of governing, but it did so in the institutionalized forms of independent, neutral colleague associations, controlling recruitment and training, providing codes of conduct and procedures of discipline...underwritten by government in the form of official recognition of license"* (Johnson 1994: 144).

The restrictive practices of professionalism's exclusive club mentality may seem on the surface to be oppositional to liberalisms free market political philosophy. But in reality, club governance formed an essential part of the emergence of liberal governmentality. Indeed, the gentleman's club was a hot bed of commerce in Victorian society, and similar to the professional club, it had clear ideas about who should get through the front door (Moran 2004). Consequently, the establishment of state-sanctioned jurisdictions for emergent professions such as medicine over the surveillance, classification and care of poor-rich, sick-healthy and mad-sane subjects was not solely the result of successful occupational strategies of advancement based upon claims to possess esoteric expertise and an altruistic code of conduct. Rather, class and gender inequalities influenced the club rule form that professional self-regulatory institutions such as the GMC took. While the establishment of the jurisdictions of emergent professions over particular elements of the general population was an outcome of programmes and policies that sought the legitimate expansion of the ability of government to shape and enhance the self-regulating capabilities of individuals along predetermined lines (Rose 1992, 1999; Barry et al. 1996). This was because *'this form of power cannot be exercised without knowing the inside of people's minds, without exploring their souls, without making them reveal their innermost secrets. It implies a knowledge of the conscience and an ability to direct it'* (Foucault 1982: 783). For example, over the last century, the medical profession has significantly contributed to an intensification in the surveillance and control of the body and the self of the individual subject under the banner of maintaining the health of the population. As Armstrong (1983: 112) notes, *'[in] the twentieth century the human body has been subjected to a more complex, yet perhaps more efficient, machinery of power which, from the moment of birth (or, more correctly, from the time of registration at an ante-natal clinic) to death, has constructed a web of investigation, observation and recording around individual bodies,, their relationships and their subjectivity, in the name of health'*.

The Enterprise Self and Neo-liberal Governmentality

The focus of governmentality is upon the role of professional expertise as a socio-technical device through which the self-regulating subjectivity of citizens is surveyed and managed at a distance. It does not view contemporary challenges to professional privilege, such as reforms to the GMC, as being only concerned with reducing the high level of freedom from outside control that the professions have historically possessed (Johnson 1994, 1995). It highlights how they are also concerned with the ultimate object of governmentality: the population in general and the individual subject-citizen in particular. Indeed, as Johnson (1994: 149) notes, *'government-initiated change has, in recent reforms, been securely linked with the political commitment to the 'sovereign consumer'. In the case of reform in the National Health Service, this translates...[to a] stress on prevention, the obligation to care for the self by adopting a healthy lifestyle, the commitment – shared with the new GP – to community care'*. From a Foucauldian perspective, contemporary sociological analyses of the professions and the governance of professional expertise must be set against a background of the re-emergence of liberalisms' possessive individualism in the last three decades in form of the enterprise self of neo-liberalist governmentality (Rose 1996a, 1999). Here, as Rose (1999: 87) notes, *'a new relation of individuals to expertise is established, based not upon welfare bureaucracies, social obligations and the inculcation of authoritatively established norms, but upon the mechanisms of the market and the imperatives of self-realization'*.

Rhodes (1994) notes that the attitude of UK government towards professionals since the 1980s has been dominated by a concern for the '3 Es': economy, efficiency and effectiveness. This is demonstrated by the rise of new public managerialism in the public sector. The growth of a rationalistic-bureaucratic managerial discourse of outcomes-based transparent standard setting and performance appraisal in the health- and social-care arena is bound up with this. Indeed, it is often argued that a *'new commercialized professionalism'* (Hanlon 1998: 54) has emerged which stresses the need for professionals to develop managerial and entrepreneurial skills (Hanlon 1994, 1998). This is a result of government attempts to improve trust in public sector services by seeking to redefine professionalism so that it becomes more commercially aware, target focused and managerially accountable.

In this regard, the Foucauldian viewpoint shares much in common with neo-Weberian accounts of the contemporary situation faced by the medical profession. Both hold that there has been an increase in government intervention in the public sector and in turn the work of health- and social-care professionals. As part of this, doctors have increasingly become entwined in a particular system of governance, namely, clinical governance, which requires the medical profession work alongside management to align clinical authority with economic viability (Flynn 2002). However, authors working from a governmentality viewpoint also see the emergence of a more economically aware form of professionalism as taking place against a background of a profound shift in *'the nature of the present'* (Rose 1992: 161) and the way *'[we] come to recognize ourselves and act upon ourselves as certain kinds of subject'* (Rose 1992: 161). For in this way, *'a person's relation to all his or her activities, and indeed his or her self, is...given the ethos and structure of the enterprise form'* (Rose 1999: 138). Certainly, from a governmentality perspective, challenges to medical autonomy in the National Health Service and the principle of medical self-regulation in the form of the GMC form part of a broader shift in the grounds under which the legitimate governance of the population is practised (Rose 1999). This is due in no small part to the ascendancy of the enterprise self throughout all spheres of modern social life. As Burchell (1993: 275) argues, *'one might want to say that the generalization of an "enterprise form" to all forms of conduct – to the conduct of organizations hitherto seen as being non-economic, to the conduct of government, and to the conduct of individuals themselves – constitutes the essential characteristic of this style of government: the promotion of an enterprise culture'*.

The emergence of neo-liberalism in the 1970s reactivated classical liberalism's concern with the liberty of the individual, advocacy of free markets and call for less direct government. It emphasised the entrepreneurial individual, endowed with freedom and autonomy, and a self-reliant ability to care for herself, driven by the desire to optimise the worth of her own existence (Rose 1993, 1996a, 1999). For example, the conservative home secretary, Douglas Hurd, stated in 1989 *'the idea of active citizenship is a necessary complement to that of the enterprise culture'* (quoted in Barnett 1991: 9). Neo-liberal forms of governmentality seek to govern through the autonomy of the governed. Citizens should be active, not passive, and democratic government must engage the self-regulating capabilities of individuals. Neo-liberal government focuses upon the use of the technologies of the self because the *'citizens of liberal democracy are to regulate themselves; government mechanisms construe them as active participants in their lives...Such a citizen subject is not to be dominated in the interests of power, but to be educated and solicited into a kind of alliance between personal objectives and ambitions and institutionally or socially prized goals or activities. Citizens shape their lives through the choices they make about family life, work, leisure, lifestyle, and personality and its expression.*

Government works by "acting at a distance" upon these choices, forging a symmetry between the attempts of individuals to make life worthwhile for themselves, and the political values of consumption, profitability, efficiency and social order' (Rose 1990: 10).

Burchell (1996: 28–29) argues that neo-liberalisms dual advocacy of the self-regulating free individual and the free market have led to *'the generalization of an "enterprise form" to all forms of conduct'*. Similarly, du Guy (1996a, b) argues that enterprise, with its focus upon energy, drive, initiative, self-reliance and personal responsibility, has assumed a near-hegemonic position in the construction of individual identities as well as in the government of organisational and everyday life. Enterprise, he argues, has assumed *'an ontological priority'* (du Guy 1996a: 181). du Guy holds that *'a discourse of enterprise makes up the individual as a particular sort of person – as an "entrepreneur of the self"'* (du Guy 1996b: 11), so *'every individual life is, in effect, structured as an enterprise of the self which each person must take responsibility for managing to their own best advantage'* (du Guy 1996b: 14). The concept of the self as enterprise requires that the possession of an essential core self is taken as the central feature of personal identity. How else could individuals be expected to become responsible for themselves and the care of their bodies and not a burden on the state? The very notion of the enterprise self requires a political commitment to the idea that all individuals are capable of self-fulfilment. This is the core mechanism by which the self-regulatory capabilities of the individual can be enhanced and entwined with the key objectives of governance: the security, health, wealth and happiness of the general population. Consequently, failure to achieve the goal of self-fulfilment is not associated with the possession of a false idea of what it means to be human, or that individuals do not possess an essential core self which is the real and true them for all eternity. Rather, it is the fault of poor choices, a lack of education or the dependency culture created by the welfare state. It is the result of learned helplessness, which in itself can be resolved with *'programmes of empowerment to enable [the individual] to assume their rightful place as self-actualizing and demanding subjects of an "advanced" liberal democracy'* (Rose 1996a: 60)

Expert Enclosures and Technologies of Performance and Agency

The ascendancy of the enterprise self has increasingly led, as Rose (1993: 285) notes, expertise being increasingly *'governed by the rationalities of competition, accountability and consumer demand'*. Rose (1993, 1996a, 1999) argues that the increasing institutionalisation of expertise during the nineteenth and twentieth centuries led to expert knowledge becoming integral to the exercise of political authority. Experts gained *'the capacity to generate "enclosures", relatively bounded locales or fields of judgment within which their authority [was] concentrated, intensified and rendered difficult to countermand'* (Rose 1996a: 50). However, as a result of the rise of the enterprise self, the enclosures are now being *'penetrated by a range of new techniques for exercising critical scrutiny over authority – budget disciplines, accountancy and audit being the three most salient'* (Rose 1996a: 54). As Osborne (1993) discusses, ever since the re-emergence of liberalism in the 1970s, there has been a gradual reformulation of health-care governance so that the field of medicine is, to a greater degree than ever before, simultaneously both governed and self-governing. In an attempt to promote public trust in public sector reforms, the state becomes increasingly involved in medical governance and seeks to promote greater accountability and transparency within professional systems of self-governance. To achieve these goals, it adopts a strategy whereby professional expertise is increasingly subjected to an additional layer of management and new formal calculative regimes (Rose and Miller 1992), including the setting of performance indicators, competency frameworks and indicative budget targets.

Rose (1996a, 1999) emphasises the enormous impact of the trend in all spheres of contemporary social life towards audit in all its guises, with its economic concern with transparent accountability and standardisation, particularly for judging the activities of experts. This is because two technologies are central to the promotion of the enterprise self at the organisational and individual levels: a technology of agency, which seeks to promote the agency, liberty and choices of the individual as they strive for personal fulfilment, and a technology of performance, which seeks to set norms, standards, benchmarks, performance indicators, quality controls and best practice standards. These help to survey, measure and render calculable the performance of individuals and organisational structures. Dean (1999: 173) notes: *'from the perspective of advanced liberal regimes of government, we witness the utilization of two distinct, yet entwined technologies: technologies of agency, which seek to enhance and improve our capabilities for participation, agreement and action, and technologies of performance, in which these capabilities are made calculable and comparable so that they might be optimized. If the former allow the transmission of flows of information from the bottom, and the formation of more or less durable identities, agencies and wills, the later make possible the indirect regulation and surveillance of these entities. These two technologies are part of a strategy in which our moral and political conduct is put into play within systems of governmental purposes'.*

Bound up with the technologies of agency and performance of the enterprise, culture is what can be called a progressive and insipid process of contractualisation. Institutional roles and social relations between individuals are increasingly defined in terms of explicit contract, or at the very least, in a contract like way. The promotion of the enterprise form involves the creation of processes where subjects and their activities are *'reconceptualized along economic lines'* (Rose 1999: 141) Indeed, Gordon (1991: 43) argues that entrepreneurial forms of governance rely on contractualisation as they seek *'the progressive enlargement of the territory of economic theory by a series of redefinitions of its object'*. Entrepreneurial forms of governance reimagine the social sphere as a form of economic activity by contractually: (a) reducing individual and institutional relationships, functions and activities to distinct units; (b) assigning clear standards and lines of accountability for the efficient performance of these units; and (c) demanding individual actors assume active responsibility for meeting performance goals, primarily by using tools such as audit, performance appraisal and performance-related pay (du Guy 1996c). Here, judgement and calculation are increasingly undertaken in economic cost-benefit terms, which give rise to what Lyotard (1984: 46) terms the performativity principle: the performances of individual subjects and organisations serve as measures of productivity or output, or displays of quality, and *'an equation between wealth, efficacy and truth is thus established'* (1984: 46).

Neo-liberal governmentality is concerned with the conduct of conduct and seeks to govern through the autonomy of the governed. It is concerned with the practices of liberty, that is, it is concerned with practices which structure, shape, predict and make calculable the operation of freedom. This is also why the traditional closed shop form of professionalism as a self-regulatory strategy for institutionalising expertise has been challenged by the rise of the enterprise self but self-regulatory privileges have not been completely erased. For in a very real and practical way, government depends upon expertise to render social realities governable – whether it is in the field of health, education or law – through shaping the self-regulating capacity of subjectivity amongst all citizens, *including* that belonging to professionals themselves. The effectiveness of neo-liberal mentalities of rule lies in their ability to align *'the objectives of authorities wishing to govern and the personal projects of those organizations, groups and individuals who are the subjects of government'* (Rose 1999: 48). The diagnostic truths and recommendations for action provided by experts such doctors play a vital role in fostering such alignments. Consequently, from this viewpoint, the emergence of a new more publically accountable form of professionalism within professions such as medicine is inextricably bound up with a broader shift in the nature and scope of legitimate authority and forms of governance within modern democratic liberal society.

Certainly, the apparent re-appropriation of rationalistic-bureaucratic technologies of performance by occupational elites within professions such as medicine appear to belong to a particular mentality of rule which recognises that *'[rule] "at a distance" [only] becomes possible when each [agent] can translate the values of others into its own terms [so] that they promote norms and standards for their own ambitions, judgments and conduc*t' Rose (1999: 50). For example, a study analysing clinical governance in primary care through the lens of governmentality by Sheaff et al. (2004) found that there had been a shift towards more formal networks of collective peer review of individual doctors' work practice, primarily through the contractual use of rationalistic-bureaucratic technologies of performance such as medical audit and evidence-based medicine. Greater peer surveillance of clinical practice is justified by doctors themselves because they sought to reduce unnecessary variations in the quality of clinical care as well as minimise economic costs. This is in addition to forestalling a perceived growing threat of managerial control over medical work. The overarching outcome of this shift towards greater surveillance and control of individual doctor's activities was an increase in the view that individual doctors could and indeed should be placed under peer surveillance and control. Hence, governing objectives for greater cost and risk containment existed alongside an increase in professional accountability and were aligned with the medical profession's self-image of itself as an independent and yet morally and socially responsible occupation.

Critical Reflections on the Governmentality Perspective

The governmentality perspective makes a significant contribution to the sociological study of the professions and professional regulation. It highlights the key role professions, such as medicine, have played in the governance of the population. In doing so, it adopts a similar critical view of the emergence of professionalism as a form of regulatory control as the neo-Weberian perspective. Importantly, it reinforces the need for current debate surrounding recent challenges to the principle of professional self-regulation, to also consider the changing nature of the relationship between subject-citizens and the state, as a result of the political and economic re-emergence of liberalism since the mid- to late 1970s. For the governmentality perspective notes the ascendancy of the concept of the enterprise self into all spheres of contemporary life. In doing so, it highlights how challenges to the principle of professional self-regulation and concurrent calls to reform the GMC can be seen to be directed towards the object of governmentality – the population in general and the individual subject-citizen in particular – as much as they are the medical profession, for medicine, and indeed the health- and social-care professions as a whole, forms but one part of a complex array of governing calculations, strategies and tactics which seek to promote the security, wealth, health and happiness of the population (Rose and Miller 1992). It is important for social scientists to recognise this. As in terms of Isaiah Berlin's (1969) famous dichotomy of positive and negative liberty, although liberal mentalities may appear at first to promote negative liberty (i.e. the personal freedom of the individual subject to decide who they are and discover what they want to be), they in reality promote positive liberty (i.e. a view of who and what a citizen-subject is and should be). This carries with it the very real danger of authoritarianism (Dumm 1996).

The governmentally perspective reinforces that modern government must seek to govern through the freedom and aspirations of their citizen-subjects so that they come to recognise and self-regulate their activities in such a way that they naturally align with broader social, economic and political objectives. This requirement has led to a critical reconfiguration of the legitimate grounds on which good governance can be practised. Hence, the field of medicine becomes more than ever before simultaneously governed and self-governing as a consequence (Osborne 1993). As illustrated by the re-appropriation by medical elites of an emergent rationalistic-bureaucratic discourse of outcomes-based standard setting and performance appraisal in the face of its increasing use by outsiders, such as hospital management, to monitor the activities of doctors. Yet a key problem with the governmentality position is that it possesses a tendency to overstate the dominance of the enterprise form in the social, economic and political spheres, as well as the construction of human subjectivity, and so an individual's sense of personal identity. Put simply, enterprise is just one of many sources which an individual can draw on to construct a sense of self.

Additionally, the governmentality perspective also tends to overstate the closeness of the relationship between professions such as medicine and the state. Certainly, it is important to recognise the close relationship that exists between the professions and the state. Indeed, in many ways, modern forms of expertise such as medicine are instruments through which state power and control over populations and individuals is enacted. Yet it is often necessary for practical reasons to demarcate professional expertise from the governing apparatus of the state. Particularly, when analysing empirically the effects of contemporary reforms on the principle of professional self-regulation, it is the neo-Weberian social closure perspective that is most useful here. Its account of the historical development of medicine's exclusive cognitive identity, which underpins its member's only stance concerning the issue of who should regulate doctors, reflects the nature of the regulatory context and the predominant form of the occupational culture of the medical profession, at least in the Anglo-American context. As illustrated by the continued insistence by medical elites that whatever changes to the current system are introduced, they should be heavily involved in deciding if a doctor is indeed fit to practise and should remain on the state-approved register of practitioners. Indeed, Kuhlmann (2006a: 222) argues doctors have been able to '*amalgamate managerialism and professionalism.....and [been able to] outflank tighter public control and [attempts] to create a comprehensive system of accountability*'. This is a point we will return too when we look at the revalidation thesis in this and subsequent chapters.

Given that the neo-Weberian perspective reflects the reality of the regulatory context, as well as offers important insights into the occupational culture of the medical profession, it is arguable that the governmentality viewpoint can and should be used to supplement and expand on the neo-Weberian perspective. Indeed, synthesising these two viewpoints is a fruitful approach to adopt when analysing the regulation of medical expertise. Both argue that it is necessary to adopt a critical and historical approach when studying how professionalism operates as a regulatory strategy, as well as when exploring the reasons behind current reforms in the regulation of professional expertise. Importantly, both hold that recent challenges to the principle of professional regulation have caused professional elites to place rank-and-file practitioners under greater surveillance and control, as they seek to maintain collective self-regulatory privileges. Here, it should be remembered that the chapter noted how the governmentality perspective argues that the field of medicine has become more than ever before simultaneously governed and self-governing (Osborne 1993). This is a state of affairs conceptualised by the neo-Weberian perspective under the banner of the restratification thesis (Freidson 1985, 1994, 2001). This thesis lies at the centre of the analysis of revalidation in Chap. 6 and reforms in the hearing of fitness to practise cases outlined in Chap. 7. Indeed, it is a key argument of this book that the restratification thesis captures and explains the causes and consequences of contemporary developments in the regulation of the medical profession. Consequently, it is vitally important in the remainder of this chapter to outline its development and key principles.

The Proletarianisation and Deprofessionalisation Theses

The restratification thesis first emerged in the mid-1980s in response to the growing recognition within sociology that something was happening to medical autonomy. It was conceptualised as being in decline by what are, respectively, called the proletarianisation and deprofessionalisation theses (Freidson 1985). Writing at the beginning of the 1970s, Haug (1973), the originator of the deprofessionalisation thesis, argued that medical autonomy was being challenged due to a process of rationalisation and codification of medical knowledge and expertise into standardised rules and procedures. She focused on the role this played in reducing the knowledge gap between patient and doctor, as well as in supporting the rejection of professional paternalism, as a more informed and critical general public became less inclined to act deferentially towards experts. Haug (1973:206–207) noted that this process was only just starting: *'[The] tension between the public demand for accountability and the professionals insistence on final authority has not yet erupted into general warfare...But there have been skirmishes'*. Haug (1973) argued that a tipping point had been reached, with medicine starting to lose its prestigious social-political position. She cited five interrelated factors to support her viewpoint. First, while medical knowledge was rapidly expanding, it was undergoing a process of codification at a general level. This, Haug argued, was leading to medicine losing its control over its defined body of knowledge due to a rise in automated retrieval systems, such as computer algorithms, for symptom assessment.

Second, the public were becoming more educated, better informed about health matters, and more likely to challenge physician authority than ever before. Third, as medical knowledge expanded, medicine as a profession was increasingly fragmenting into specialties and sub-specialties, with individual doctors becoming ever more dependent upon each other for expert advice, as well as ever more dependent upon non-medical expertise. One physician no longer held all the power over a patient. This reduced even further individual and collective autonomy. Fourth, there had been a growth in the patient self-help groups and a rise in alternative medicine as public trust and belief in medical expertise declined. It became ever clearer through high-profile media cases that in reality, medicine's cognitive and altruistic claims did not live up to expectation. Fifth, increases in medical care costs meant the public were demanding doctors be held more accountable for their actions. Indeed, in some cases, they wanted the principle of medical self-regulation to be abolished.

The deprofessionalisation thesis tends to focus on topics that indicate that there has been a decline in public trust of medicine and the threat this poses to the principle of professional self-regulation. The growth of media coverage of gross medical malpractice cases, like the Shipman case discussed in Chap. 3, is a good example of this process. It focuses upon the fact that attitudes to traditional forms of authority are changing and highlights that the public increasingly expects their governing institutions to operate in a transparent and accountable manner. In contrast, the proletarianisation thesis highlights the existence of the potential for expert work in general, and medical work in particular, to become subject to rationalisation and routinisation. Hence, today's indeterminacy is becoming tomorrow's technicality. It focuses upon how this causes medical work to become subject to managerial bureaucratic control in the name of controlling costs and promoting consumer choice.

Writing at the same time as Haug (1973), the originator of the proletarianisation thesis, Oppenheimer (1973), held that the work of professionals was becoming subject to a process of rationalisation in the name of economy and efficacy. This had happened in the factory at the beginning of the industrial revolution over 150 years earlier. Like Haug (1973), Oppenheimer held that the scientific nature of modern specialist knowledge and expertise meant it was open to communication as a set of rules, procedures and operational imperatives where passed on to others who had not received any formal training. Work tasks could be broken down into parts so that, on one hand, workers performed one or a handful of tasks from a whole process, and on other hand, administrative and bureaucratic authorities could determine overall working conditions and priorities. Furthermore, Oppenheimer focused upon the fact that professionals were operating in large organisational settings (such as modern hospitals) as salaried employees, where he held the growth of bureaucratic rules, procedures and authority was undermining professional autonomy. Oppenheimer (1973: 214) notes that *'the bureaucratized workplace....[tends to replace]...in the professionals' workplace factory-like conditions – there are fixed jurisdictions, ordered by rules established by others; there is a hierarchical command system; jobs are entered and mobility exists on the basis of performance in uniform tasks, examinations, or the achievement of certification, or "degrees"...The gap between what the worker does and the end product, increases'*.

For Oppenheimer, a process began whereby administrative routines, measures and targets controlled professional work. His central thesis was that the work of professionals was increasingly becoming subordinated within bureaucratic structures to the control of administrative elites operating under fixed rules and procedures, which the professions had no control over. McKinlay (1977), McKinlay and Arches (1985), and McKinlay and Stoeckle (1988) noticed this theme from an explicit neo-Marxist perspective and in the context of medicine. They held that as medicine had advanced and entered large-scale corporate and bureaucratic settings, physicians lost several professional prerogatives associated with the principle of self-regulation, such as control over entrance criteria, training context and content, workplace autonomy and the object, tools and means and remuneration of their labour. They discussed how the American federal government and managerial corporate rationalisers were affecting the content of medical work and medical school curricula. Medicine was becoming fragmented into sub-specialisms, as medical knowledge expanded. Non-medical staff that operated largely outside of direct medical control were also intervening in the doctor-patient relationship as medical techniques became ever more reliant on new technologies. Patients were increasingly the clients of the organisations doctors worked within, instead of being the direct responsibility of an individual doctor. Under these circumstances, they felt medicine as an occupation could no longer be held to be professionally dominant. Contrasting the position of self-employed physicians at the turn of the twentieth century with their modern-day counterparts, McKinlay and Stoeckle (1988: 201) concluded that *'[every] single prerogative listed has changed, many changes occurring over the last decade. The net effect of the erosion of these prerogatives is the reduction of the members of a professional group to some common level in the service of the broader interests of capital accumulation'*. They argued that while the proletariat possess a false consciousness regarding their true exploited position in capitalist society, doctors similarly possessed a false consciousness with regards to their true social position: *'For doctors who are increasingly subject to this process, it is masked by their false consciousness concerning the significance of their everyday activities and by an elitist conception of their role so that even if the process is recognized, doctors are quite reluctant to admit it'* McKinlay and Stoeckle (1988: 201).

McKinlay and Stoeckle's discussion of the proletarianisation thesis fails to recognise that doctors are not quite like other workers. It is highly questionable that by the 1980s, the entire labour force in the Anglo-American context had been progressively proletarianised under advanced capitalism. Regardless of their salaried status and managerial inroads into controlling medical work, doctors retained the power to direct and supervise the work of others and maintain a range of specialist skills, which enabled them to collectively bargain for positions of high social privilege, status and power. Neo-Marxists like Navarro (1988) who are critical of the work of Freidson admit he is correct in maintaining that medical autonomy is essentially a collective not individual property. Furthermore, relative to other health-care occupations, medicine occupies a prominent position in the health-care arena as no other occupation has the capacity to dominate it. This being said, by the mid-1980s, Freidson (1994) recognised that medicine was, first, coming under pressure from the state to reform its regulatory and training institutions, second, was being placed under third-party greater surveillance and control by the rise of managerialism and, third, was no longer as dominant over other occupations in the health-care arena as it once had been. Nursing, for example, was establishing its independence from traditional medical control as the state started to emphasise multidisciplinary working patterns in its attempts to reduce costs, maximise efficiency and respond to a rise of consumerist calls for increased patient choice. It is from this position that the restratification thesis emerged (Freidson 1985, 1994).

The Restratification Thesis

In the 1970s, critical commentators shared a common emphasis on viewing professionalism ideologically as an exclusionary self-regulatory strategy for organising the performance of professional work. As Chap. 4 discussed, this revolves around the principle that members of a profession must exercise control over their work and the standards by which work outcomes are judged, due to the specialist nature of their expertise. Occupational control over members training and discipline forms a logical part of this viewpoint. Yet, by the mid-1980s, the changes posited by the respective deprofessionalisation and proletarianisation theses were acknowledged as actually starting to occur by sociologists (Coburn and Willis 2000). It was beginning to look like medical dominance and autonomy was going into long-term decline, just as the deprofessionalisation and proletarianisation theses had predicted. Rapid advances in medical knowledge made it apparent that medicine was becoming less homogenous and fragmenting into sub-specialties, as new diagnostic and therapeutic technologies developed due to the advent of the computer age and advances in pharmacology, molecular biology, genetics and immunology (Gabe et al. 1994). This caused medicine to become ever more dependent upon non-medical occupations operating outside of its direct jurisdiction to treat illness and disease (Elston 1997).

Concurrent with the rapid growth in medical expertise and the growing internal fragmentation of the profession was a rise in managerial attempts to control medical work. There was the ascendancy of managers or corporate rationalisers as the state sought to contain burgeoning health-care costs (Coburn and Willis 2000). The invasion of the state via management into medical turf was also related to growing public concern with the risks involved in modern medical treatment. High-profile media cases engendered doubts in the consciousness of the public concerning the ability of medicine to ensure individual doctors possessed high ethical standards. They also contributed further to an already burgeoning consumerist demand for greater patient choice and control over medical encounters as well as health-care organisation and delivery. This was reflected in the growth of alternative medicine, an increase in the threat of patient complaints and medical litigation, as well as the presence of a high level of dissatisfaction amongst patients with the doctor's communication and information sharing skills.

Despite these broad changes, the proletarianisation and deprofessionalisation theses and their applicability outside the United States of America was questioned (Gabe et al. 1991). By the end of the 1980s, many commentators agreed that although clear differences between the American and UK health-care systems remained, *'[both] countries are moving towards greater third-party control of both global health care budgets and clinical decisions'* (Harrison and Schultz 1989). Yet as Elston (1991) pointed out, although there had been a rise in managerialism on both sides of the Atlantic, after the establishment of the NHS in 1948, the majority of UK doctors had become salaried state employees. This meant greater potential existed for direct state interference with regards to medical autonomy in the NHS, as well as medical school admission numbers and curricula content to meet NHS workforce planning needs. Elston (1991: 66) highlighted that under the conditions assumed by McKinlay and Arches (1985), *'the proletarianization of the British medical profession was virtually completed forty years ago'*. Rather than showing that UK medicine had been proletarianised earlier than American medicine, Elston (1991: 66) argued that this demonstrated the importance of *'disaggregating components of autonomy in analysis'*. Elston (1991: 61) defined 'medical dominance' as medicine's authority over others and subdivided it into (1) 'social authority', which related to medical control over the actions of others, and (2) 'cultural authority', which related to the acceptance of medical definitions of reality and therefore medical judgements being accepted as valid and true. She divided 'medical autonomy' into three main categories: (1) 'economic autonomy' (the right of doctors to determine their remuneration), (2) 'political autonomy' (the right of doctors to make policy decisions as the legitimate experts on health matters) and (3) 'clinical or technical autonomy' (the right of the profession to set its own standards and control clinical performance, as exercised through clinical freedom at the bedside and collegial control over recruitment, training and discipline) (Elston 1991:61).

Elston argued that it was possible for the different components of medical autonomy to operate independently from each other. She held that her historical and comparative analysis of UK and American medicine showed that for much of the twentieth century, American doctors enjoyed a considerably higher level of 'economic autonomy' compared to their UK counterparts. The reduction of UK doctors' 'economic autonomy' because of their employment in the NHS after its inception in 1948 had not affected their other professional privileges. The profession's prominent position within the NHS meant it possessed considerable 'social authority' over other occupations operating in the NHS. Likewise, its 'clinical autonomy' and 'political autonomy' were enhanced, as it exerted a powerful influence over the shaping of health-care policy and practice. Approximately 80% of all health expenditure was determined by decisions made by doctors with the government leaving medicine effectively in charge of the NHS during this golden age of medical power (Klein 1983).

Elston (1991) was equally critical of the deprofessionalisation thesis. Focusing on doctor's right to self-regulate their activities, she acknowledged that there had been challenges to the principle of medical self-regulation, as embodied by the GMC. Nevertheless, she argued, *'the modifications of professional self-regulation appear as a series of incremental adjustments to contain criticism rather than substantial diminution of collegiate control'* (Elston 1991: 81). Furthermore, she recognised that patients demanded to be given greater informed choice, and this was tied up with the dominance of neo-liberal economic ideology within the health-care arena and specifically NHS reforms emphasising *'the discipline of the market and consumer power'* (Elston 1991: 78). However, she found little hard evidence to support the viewpoint that the public was rejecting the validity of medical science's 'cultural authority'. She acknowledged that the potential power of medical knowledge and expertise to cure all ills was increasingly questioned. But she argued that *'[the] growth of the women's self-help movement and holistic well-women centres and the apparently increasing use of alternative practitioners suggests some of the disillusioned are exiting the system, but only partially and on a small scale... [furthermore there]... is little baseline data against which changes in the level of public confidence in and valuation of medicine can be tested'* (Elston 1991: 82).

Additionally, sociologists could not agree if the proletarianisation and deprofessionalisation theses were applicable within the American context, where they were first generated. Freidson was a key critic (i.e. Freidson 1985, 1994). He agreed that changes were occurring in medicine's relationship with the public and acknowledged that this was due to medical knowledge and expertise expanding, as well as becoming formalised into rules and procedures with the advent of computer technology and the information and communication revolutions. However, he argued that *'the professions...continue to possess a monopoly over at least some important segment of formal knowledge that does not shrink over time, even though both competitors and rising levels of lay knowledge may nibble away at the edges. New knowledge is constantly acquired that takes the place of what has been lost and thereby maintains the knowledge gap. Similarly, while the power of computer technology in storing codified knowledge cannot be ignored, it is the members of each profession who determine what is to be stored and how it is to be done, and who are equipped to interpret and employ what is retrieved effectively. With a continual knowledge gap, potentially universal access to stored data is meaningless. In sum, while the events highlighted by proponents of the deprofessionalization thesis are important, the argument that members of the professions are losing their relative prestige and respect, their special expertise, or their monopoly over the exercise of that expertise over time are not persuasive'* (Freidson 1994: 134–135).

Although he recognised that medical paternalism had been rejected and the public were more active health-care consumers, Freidson dismissed the idea of deprofessionalisation as he held that medicine was not losing control of its monopoly over its expertise. He believed that the development of new techniques to monitor the efficiency of performance and the allocation of resources did not in itself reduce medical autonomy. What matters is whose criteria for evaluation are used and who controls any ensuing action. This is an important point, for as the chapter has repeatedly highlighted, to function ideologically as a method of occupational control, professionalism requires that occupational members control the technical evaluation of work activities. In the context of the proletarianisation thesis, the growing threat of bureaucratic managerial control over medical work does challenge medical professionalism as it can, for example, introduce non-medical criteria from which to judge work performance. Freidson recognised this. Furthermore, he held that many of the changes that neo-Marxist authors such as McKinlay and Stoeckle (1988) identified have indeed occurred. It was true, for instance, that in America, a large number of doctors had moved from possessing self-employed to employed status. Concurrent with this shift were moves towards subjecting the work of individual doctors to performance evaluation and management control (Coburn and Willis 2000). This was because the spread of managed care across America to control costs and improve efficiency had created strong pressures to reduce medical autonomy in clinical decision-making.

However, Freidson retorted that while the individual autonomy of doctors was affected by this state of affairs, the collective institutional autonomy of the profession as a whole remained intact (Freidson 1994). This was because concurrent with these, changes to the health-care system in America had been the growth in co-opted medically qualified managers controlling the surveillance and evaluation of medical work. Freidson argued that the rise of medically qualified mangers illustrated that medicine was not undergoing a process of proletarianisation, but rather was dividing into more pronounced elite and rank-and-file segments. For him, it became internally fragmented due to advances in medical knowledge as well as the threat played by the rise of managerialism on one hand and the consumerist 'patient choice' movement on the other: *'Professionalism is being reborn in a hierarchical form in which everyday practitioners become subject to the control of professional elites who continue to exercise the considerable technical, administrative, and cultural authority that the professions have had in the past'* (Freidson 1994: 9).

Freidson believed that the loyalties of these co-opted doctors ultimately lay with their clinical colleagues, not their corporate masters. He held that the purpose of elite placing the rank and file under ever more formal surveillance and control was to maintain collective privileges and sustain medical professionalism as a methodology of occupational control: *'[These] changes do not affect the position of the profession as a corporate body...so much as they affect the internal organization of the profession in the relation amongst physicians. In essence, I suggest, they are creating more distinct and formal patterns of stratification within the profession than have existed in the past, with the position of the rank and file practitioner changing most markedly'* (Freidson 1985: 6). For Freidson, the rise of peer-review mechanisms brought about by the increased use of surveillance tools such as medical audit was not a sign of the proletarianisation of medicine. Rather they were an essential part of the process of restratification, which he held was occurring within medicine. Audit and peer review were well-established surveillance mechanisms in America by the late 1970s, unlike in the UK (Harrison and Schulz 1989). Furthermore, Freidson (1994: 145) went on to argue that *'there is little evidence that the special status of rank and file professionals will deteriorate so much that they will find themselves in the same position as other workers. Even though they will be subject to more formal controls than in the past... [in] all likelihood, they will also exercise considerably more discretion than other workers in performing their work, and will be able to participate in formulating standards and evaluating their own performance through some type of peer review. Finally they will still enjoy at least occupational kinship with their superiors'*.

Freidson's arguments for the existence of restratification within medicine feel like good common sense. Because any attempt to raise standards and cut costs clearly requires the cooperation of the medical profession (Gray and Harrison 2004). Also, it is somewhat ironic that at the same time, sociologists were arguing about a possible decline in the status and power of the medical profession and modern medical technology and expertise were making significant improvements to people's lives. As Kelly and Field (1994: 36) note, *'to deny the effectiveness of modern medical procedures such as coronary artery bypass, renal dialysis, hip replacement, cataract surgery, blood transfusion, the pharmacology of pain relief and the routine control of physical symptoms in restoring or improving the quality of life for those suffering from chronic illness is to deny the validity of the everyday experiences of the lay public in modern Britain. In stressing the limitations and costs of medical interventions, the physical and social contributions of modern medicine are all too frequently ignored'*.

Medicine and Managerial Corporate Rationalisers Revisited

Elston (1991) argued following Freidson (1994) that medicine in the UK was undergoing a process of restratification. She noted that the fact it was already embedded within the managerial and bureaucratic structures of the NHS facilitated this process. She agreed that managerial corporate rationalisers were in ascendancy. Not least of all because state reforms to health care throughout the 1980s had introduced performance surveillance mechanisms into the NHS, such as indicative clinical budgets, prescribing lists and medical audit. However, she argued that the embedded nature of medical expertise in the NHS, as well as the political bargaining of professional elites such as the royal colleges, meant that these mechanisms were largely placed in the hands of co-opted medical-managers. Elston's advocacy of Freidson's restratification thesis is supported by empirical work conducted with clinical directors in the UK by Kitchener (2000) between 1991 and 1997. The role of clinical director was created because of the introduction of an internal market in the NHS in the early 1990s. This separated purchasing and providing functions in order to improve efficiency by introducing an element of open-market competition into service provision, in line with neo-liberal conservative economic policy (Elston 1991).

Clinical directors operate as part of a clinical directorate, which is under the control of a medical director who sits on the hospital board. They are responsible within particular clinical areas for overall budgets, the recruitment of staff and monitoring service quality. Kitchener (2000) interviewed a number of clinical directors over time as well as non-medical administrative hospital staff. Kitchener (2000: 149) concluded that *'little evidence emerged from this study to indicate any significant appropriation of clinical tasks or decisions by other groups....[indeed clinical directors have]...proved successful in protecting medical autonomy and resisting the increased managerial control...The result is that peer review is still widely perceived to be the primary means of quality control in UK hospitals...This position remains far removed from a managerial process of quality assurance that the reformers hoped would allow externally driven performance analysis to reduce clinical autonomy and costs'*.

Elston (1991) argued that in the UK context, the cultural authority (i.e. the belief that medical definitions of reality are valid and true) of medical judgements had remained and would remain intact. In line with Freidson's (1994) restratification thesis, she recognised that individual doctor's clinical autonomy would slowly decline, and at the same time, there would be an increase in the formalisation of the methods by which the profession's own elite institution controlled their members. Indeed, Elston (1991: 96) argued, *'it may turn out that it is the corporate rationalizers within the profession who are in the ascendant in Britain'*. Alongside Kitchener's (2000) empirical work with clinical directors, Elston's restratification arguments concerning the possible ascendancy of corporate rationalisers within UK medicine are also supported by Armstrong (2002). Armstrong argued that what he called a medical administrative elite had emerged, grouped around the academy and the professional colleges. They were concerned with standardising the everyday clinical decisions of rank-and-file doctors using evidence-based medicine. This focused upon standardising clinical judgements by disseminating the results of randomised controlled clinical trails through *'formalized tools such as audits, clinical guidelines and protocols'* (Armstrong 2002: 1772).

Random control clinical trials were used because they represent the pinnacle of medicine's cultural authority due to their apparent objective and value-neutral scientific methodology. Elston's (1991) ideas about medicine's cultural authority were also endorsed by Harrison and Ahmad (2000) whose analysis of medical autonomy in the UK was undertaken at three different levels. They held that four strands operate at the micro-level of medical autonomy: (1) 'control over diagnosis and treatment', that is, decisions regarding what tests and examinations are in order and what drugs and procedures to prescribe or who to refer a patient to; (2) 'control over evaluation of care', that is, judgements concerning the appropriateness of treatment; (3) 'control over the nature and volume of medical tasks', that is, the ability to self-manage workloads and priorities; and (4) 'contractual independence', that is, the right to engage in private practice. At the meso-level is the relationship between the state and the profession, including the legal basis of the right to self-regulation and state recognition of the British Medical Association as medicine's peak association. Finally, at the macro-level is the biomedical model. This is akin to Elston's (1991) concept of cultural authority as it relates to the social and intellectual prominence possessed by medical knowledge. It holds that the authority of medical judgements lies ultimately in their apparently scientific, objective and value-neutral nature.

Harrison and Ahmad reviewed developments in the health-care arena between 1975 and 2000. With particular attention to the rise of managerialism and greater state intervention into the principle of medical self-regulation through the establishment of *'clinical governance' bodies such as NICE. Harrison and Ahmad (*2000*:138) concluded that 'a not insignificant decline in the autonomy and dominance of British medicine has occurred over the last twenty-five years...The decline is clearest at the micro-level of clinical autonomy and at the meso-level of corporatist relations with government even though at the time of writing the Labor institutions of clinical governance and primary care organization are only just coming into existence'.* From the perspective of individual doctors, medical autonomy was in decline. They also argued, contra Elston, that the principle of medical regulation was being successfully challenged. This was because of incidents such as the Bristol case. Yet they held like Elston that *'the dominance of the 'biomedical model' at the macro level remains largely intact'* (Harrison and Ahmad 2000: 137). Furthermore, they discussed the rise of administrative elites within medicine, with royal college members and medical academics being engaged in what they called the guideline industry. They also noted that the biomedical model was being re-appropriated by non-medical management and the state as they sought to curtail medical autonomy in order to control health-care costs. Indeed, *'many of the manifestations of managerialism outlined...depend upon it: observation of medical practice variation, clinical performance indicators, the quasi-market and clinical guidelines are examples'* (Harrison and Ahmad 2000: 138).

In summary, for Harrison and Ahmed, managerial corporate rationalisers were seeking to curtail the autonomy of doctors using the outcomes generated by medicines own corporate rationalisers working in the guideline industry. While the state was adopting a rationalistic-bureaucratic discourse of performance management to justify policy changes regarding the governance of medical work in particular and the delivery of health care in general. For example, the National Institute for Clinical Excellence (NICE) now approves clinical guidelines prepared by professional and academic institutional elites. Here, Harrison and Ahmed (2000: 138) argued that there has been a rise in what they call scientific-bureaucratic medicine which is *'scientific in the sense that its prescriptions for treatment are drawn from an externally generated body of research knowledge, and bureaucratic in the sense that it is implemented through bureaucratic rules (albeit of a very specialized kind), namely, clinical guidelines'.*

Armstrong (2002) agrees with Harrison and Ahmed (2000) and Harrison and Dowswell (2002) that state-backed administrative systems are utilising medical elites to reduce clinical variation, and so eliminate risk by promoting technicality and reducing indeterminacy, through the generation of clinical guidelines and protocols via the biomedical research model. However, he cites the work of Jamous and Peloille (1970) and concludes that *'in effect, GPs can be seen as attempting to maintain "indeterminacy" in their everyday work – the traditional basis for professional status – in the face of a new forms of "technicality" promoted, ironically, by their colleagues in the medical elite'* (Armstrong 2002: 1776). In short, the work of Armstrong (2005) reinforces that the situation is perhaps not as straightforward as Harrison and Ahmed (2000) make out. Particularly in terms of the day-to-day operation of clinical autonomy in busy, fragmented and fractured clinical settings where the immediate priority of doctors and other health- and social-care professions lies with saving lives not adhering strictly to clinical protocols and formalised treatment guidelines under the ever-watchful eye of hospital (or peer) performance management mechanisms.

Furthermore, performance management tools such as audit and appraisal may well possess the potential to constrain and shape professional practice, but equally the indeterminate nature of professional judgement means practitioners are often able to subvert and even counter-colonise them (i.e. see Berg 1997; Basky 1999; Armstrong 2002). For example, the growth of co-opted medical-managers in the form of medical or clinical directors has arguably helped maintain collective quality control privileges in the face of growing NHS managerial surveillance and control of medical work (Kuhlmann and Allsop 2008). Certainly, research by Waring (2005, 2007) reinforces that medical line managers are harnessing their rank-and-file colleague's perception of threats to professional autonomy and self-regulation as a coercive means of ensuring at least the appearance of adherence to more accountable and transparent forms of medical governance in the clinical practice setting. Yet at the same time, they also show how autonomy of judgement by and large remains in operation at a day-to-day level. Chamberlain (2010), in his review of performance appraisal within medicine, notes how clinicians and hospital management collude in creative 'game playing' around performance targets and how this enables clinicians to act with a considerable degree of autonomy within clinical situations while post hoc justifying their actions within the context of clinical guidelines.

Additionally, and in the context of sociological consideration of the decline of medical autonomy and challenges to the principle of self-regulation, counter-colonisation is a strategy which to some extent protects collective self-regulatory professional privileges. For example, the growth of co-opted medical-managers in the form of medical or clinical directors has arguably helped maintain collective quality control privileges in the face of growing NHS managerial surveillance and control of medical work (Kuhlmann and Allsop 2008). Certainly, research by and Waring (2007) reinforces that medical line managers are harnessing their rank-and-file colleague's perception of threats to professional autonomy and self-regulation as a coercive means of ensuring at least the appearance of adherence to more accountable and transparent forms of medical governance in the clinical practice setting. Yet in doing so, they are protecting medical autonomy as much as they are diminishing it. This is arguably also happening within the context of the restructuring of the GMC with the royal colleges in particular playing an increasingly key role in ensuring that regulatory reforms recognise that some form of medical autonomy is needed, particularly in relation to clinical performance. As we shall see in the discussion of revalidation in Chap. 6 and the handling of fitness to practise cases in Chap. 7, it can be argued that the restratification thesis holds true when contemporary developments in these two key areas of medical regulation are explored. Yes, it is certainly true that rank-and-file doctors are coming under greater surveillance and control, but at the same time, their elite institutions continue to play a key role in securing robust training and regulatory mechanisms are in place.

Conclusion

This chapter has outlined the development of the sociological analysis of the professions over the last three decades. It has highlighted the value of synthesising the governmentality and neo-Weberian perspectives when analysing contemporary developments in the regulation of medicine. In doing so, the role of the restratification thesis has been outlined. This argues that medical autonomy has been challenged by the rise of patient's rights movements and the growth of managerialism within the health-care system. Yet it argues that in response to this, we have seen a shift within the medical profession into more pronounced elite and rank-and-file roles, with elites placing rank-and-file doctors under increased surveillance and control via performance management, as they seek to maintain medical autonomy in the form of the principle of professional self-regulation and clinical freedom at the bedside. Here, it was noted that medicine's esoteric expertise and claims to altruism continue to play a key role in maintaining medical autonomy in the face of changes in the governance of professional forms of expertise. Yes, the conditions under which medical autonomy must operate have changed. Yes, medicine has become more open and accountable and transparent in its form, content and dealings. But doctors still individually and collectively possess a great deal of autonomy and control over their affairs. For no matter how accountable they seem to become, in reality, it is the medical profession alone which possesses the necessary technical expertise to judge if a member of the medical club is underperforming. It is the further exploration of this state of affairs for the future of medical regulation which is the focus of the next three chapters. The end-of-chapter self-study activities will help you to consolidate what you have learnt before moving on to consider the contents of these chapters.

Self-Study Activity

1. Write a 1,000-word essay which critically evaluates the strengths and weaknesses of the governmentality perspective of occupational groups classified as professions and their relationship to the state.
2. Produce a 15-min PowerPoint presentation which critically considers the value of the respective deprofessionalisation and proletarianisation theses in regard to the analysis of contemporary developments in the regulation of medicine.
3. Write a 100-word essay which critically evaluates the contribution of the restratification thesis and the role played by medicine's esoteric expertise in enabling it to maintain medical autonomy.

References

Armstrong, D. (1983). *Political anatomy of the body medical knowledge in Britain in the twentieth century.* Cambridge: Cambridge University Press.

Armstrong, D. (2002). Clinical autonomy, individual and collective: The problem of changing doctors behaviour. Social Science & Medicine, 55, 1771–1777.

Armstrong, M. (2005). *Managing performance*. London: Institute of Personal development.

Barnett, M. (1991). *The politics of truth*. Cambridge: Polity Press.

Barry, A., Osborne, T., & Rose, N. (1996). *Foucault and political reason*. London: University College London Press.

Basky, G. (1999). Doctors resist adopting clinical guidelines. *British Medical Journal, 318*, 1370.

Berg, M. (1997). Problems and promises of the protocol. *Social Science & Medicine, 44*(8), 1081–1088.

Berlin, I. (1969) Two concepts of liberty. In I. Berlin (2002). *Four essays on liberty.* Oxford: Oxford University Press.

Burchell, G. (1993). Liberal government and techniques of the self. *Economy and Society, 22*(3), 267–282.

Burchell, G. (1996). Liberal government and techniques of the self. In A. Barry, T. Osborne, & N. Rose (Eds.), *Foucault and political reason* (pp. 133–167). London: University College London Press.

Burchell, G., Gordon, C., & Miller, P. (Eds.). (1991). *The Foucault effect: Studies in governmentality*. London: Harvester Wheatsheaf.

Chamberlain, J. M. (2010). Performance appraisal for doctors: A case of paperwork compliance. *Sociological Research Online, 15*(1): 1–11. Available online: http://www.socresonline.org.uk/15/1/8.html. Accessed 1 Dec 2011.
Coburn, D., & Willis, E. (2000). The medical profession: Knowledge, power and autonomy. In G. L. Albrecht, R. Fitzpatrick, & S. C. Scrimshaw (Eds.), *The handbook of social science and medicine* (pp. 155–175). London: Sage Publications.
Dean, M. (1999). *Governmentality: Power and rule in modern society*. London: Sage Publications.
Deleuze, G. (1988). *Foucault*. Minneapolis: University of Minnesota Press.
du Guy, P. (1996a). *Consumption and identity at work*. London: Sage Publications.
du Guy, P. (1996b, May). *Organising identity making people up at work*. Paper presented at the Southampton Space and Identity Seminar Series.
du Guy, P. (1996c). Organising identity: Entrepreneurial governance and public management. In S. Hall & P. du Guy (Eds.), *Questions of cultural identity* (pp. 154–174). London: Sage Publications.
Dumm, T. L. (1996). *Michel Foucault and the politics of freedom*. London: Sage Publications.
Elston, M. A. (1991). The politics of professional power: Medicine in a changing medical service. In J. Gabe, M. Calman, & M. Bury (Eds.), *The sociology of the health service* (pp. 188–203). London: Routledge.
Elston, M. A. (1997). *The sociology of medical science and technology*. Oxford: Blackwell.
Flynn, R. (2002). Clinical governance and governmentality. *Health, Risk and Society, 4*(2), 134–167.
Foucault, M. (1965). *Madness and civilisation: The history of insanity in the age of reason*. New York: Random House.
Foucault, M. (1970). *The order of things*. London: Tavistock.
Foucault, M. (1972). *The archaeology of knowledge*. London: Tavistock.
Foucault, M. (1977). *Discipline and punish: The birth of the prison*. London: Allen Lane.
Foucault, M. (1979). *The history sexuality volume 1*. London: Allen Lane.
Foucault, M. (1980). Two lectures. In C. Gordon (Ed.), *Power/knowledge selected interviews and other writings 1972: 1977* (pp. 8–28). Brighton: Harvester Press.
Foucault, M. (1982). The subject and power. *Critical Inquiry, 7*, 777–795.
Foucault, M. (1985a). *The history sexuality* (Volume 2: The use of pleasure). London: Allen Lane.
Foucault, M. (1985b). Sexuality and solitude. In M. Blonsky (Ed.), *On signs:A semiotics Reade*. London: Blackwell.
Foucault, M. (1986). *The history sexuality* (Volume 3: The care of the self). London: Allen Lane.
Foucault, M. (1989). *The birth of the clinic: An archaeology of medical perception*. London: Routledge Classics.
Foucault, M. (1991a). The ethic of care for the self as a practice of freedom: An interview. In J. Bernauer & D. Rasmussen (Eds.), *The final Foucault* (pp. 27–46). Cambridge: MIT Press.
Foucault, M. (1991b). Governmentality. In G. Burchell, C. Gordon, & P. Miller (Eds.), *The Foucault effect: Studies in governmentality* (pp. 11–27). London: Harvester Wheatsheaf.
Freidson, E. (1985). The reorganisation of the medical profession. *Medical Care Review, 42*, 1–20.
Freidson, E. (1994). *Professionalism reborn: Theory, prophecy and policy*. Cambridge: Polity Press.
Freidson, E. (2001). *Professionalism: The third logic*. Cambridge: Polity Press.
Gabe, J., Calman, M., & Bury, M. (Eds.). (1991). *The sociology of the health service*. London: Routledge.
Gabe, J., Kelleher, D., & Williams, G. (1994). *Challenging medicine*. London: Routledge.
Gordon, C. (1991). Governmental rationality: An introduction. In G. Burchell, C. Gordon, & P. Miller (Eds.), *The Foucault effect: Studies in governmentality* (pp. 134–167). London: Harvester Wheatsheaf.
Hanlon, G. (1994). *The commercialisation of accountancy*. London: Macmillan.
Hanlon, G. (1998). Professionalism as enterprise. *Sociology, 32*(1), 43–63.
Harrison, S. (2004). Governing medicine: Science and practice. In A. Gray & S. Harrison (Eds.), *Governing medicine: Theory and practice* (pp. 187–207). Milton Keynes: Open University Press.

Harrison, S., & Ahmad, W. I. U. (2000). Medical autonomy and the UK state: 1975 to 2025. *Sociology, 34*(1), 129–146.

Harrison, S., & Dowswell, G. (2002). Autonomy and bureaucratic accountability in primary care: What English general practitioners say. *Sociology of Health & Illness, 24*(2), 217–241.

Harrison, S., & Schultz, R. L. (1989). Clinical autonomy in the UK and the USA: Contrasts and convergence. In G. Freddi & J. W. Bjorkman (Eds.), *Controlling medical professionals: The comparative politics of health governance* (pp. 122–145). London: Sage Publications.

Haug, M. R. (1973). Deprofessionalization: An alternative hypothesis for the future. In P. Halmos (Ed.), *Professionalization and social change* (Sociological Review Monograph No 20, pp. 89–111). Keele: University of Keele.

Jamous, H., & Peliolle, B. (1970). Changes in the French university-hospital system. In J. A. Jackson (Ed.), *Professions and professionalization* (pp. 93–113). London: Cambridge University Press.

Johnson, T. J. (1994). Expertise and the state. In M. Gane & T. J. Johnson (Eds.), *Foucault's new domains*. London: Routledge.

Johnson, T. J. (1995). Governmentality and the institutionalization of expertise. In T. Johnson, G. Larkin, & M. Saks (Eds.), Health professions and the state in Europe. London: Routledge.

Kelly, M., & Field, D. (1994). Comments on the rejection of the biomedical model in sociological discourse. *Medical Sociology News, 19*, 34–37.

Kitchener, M. (2000). The bureaucratization of professional roles: The case of clinical directors in UK hospitals. *Organization, 7*, 129–154.

Klein, R. E. (1983 first edition and 1989 second edition). *The politics of the national health service.* London: Longman.

Kuhlmann, E. (2006a). *Modernizing health care: Reinventing professions, the state and the public.* Bristol: Policy Press.

Kuhlmann, E., & Allsop, J. (2008). Professional self-regulation in a changing architecture of governance: Comparing health policy in the UK and German. *Policy and Politics, 36*(2), 173–189.

Lyotard, J. F (1984). *The postmodern condition a report on knowledge*. Manchester: University of Manchester Press.

Macpherson, C. B. (1962). *The political theory of possessive individualism*. Oxford: Clarendon Press.

McKinlay, J. B. (1977). The business of good doctoring or doctoring as good business: Reflections on Freidson's views on the medical game. *International Journal of Health Services, 7*, 459–483.

McKinlay, J. B., & Arches, J. (1985). Towards the proletarianization of physicians. *International Journal of Health Services, 15*, 161–195.

McKinlay, J. B., & Stoeckle, J. D. (1988). Corporatization and the social transformation of doctoring. *International Journal of Health Services, 18*, 191–205.

Moran, M. (2004). Governing doctors in the British regulatory state. In A. Gray & S. Harrison (Eds.), *Governing medicine: Theory and practice* (pp. 117–137). Milton Keynes: Open University Press.

Navarro, V. (1980). Work ideology and medicine. *International Journal of Health Services, 10*, 523–550.

Navarro, V. (1988). Professional dominance or proletarianization? Neither. *Millbank Quarterly, 66*, 57–75.

Oppenheimer, M. (1973). The proletarianization of the professional. In P. Halmos (Ed.), *Professionalization and social change* (Sociological Review Monograph No 20, pp. 145–173). Keele: University of Keele.

Osborne, T. (1993). On liberalism, neo-liberalism and the liberal profession of medicine. *Economy and Society, 22*(3), 345–356.

Peters, M. (2001). *Poststructuralist, Marxism and NeoLiberalism: Between theory and practice.* London: Rowman and Littlefield.

Peterson, A., & Bunton, R. (Eds.). (1997). *Foucault health and medicine*. London: Routledge.

Rhodes, R. (1994). The hollowing out of the state: The changing nature of public services in Britain. *Political Quarterly, 65*, 138–151.

Rose, N. (1990). *Governing the soul the shaping of the private self*. London: Routledge.

Rose, N. (1992). Governing the enterprise self. In P. Heelas & P. Morris (Eds.), *The values of the enterprise culture* (pp. 56–82). London: Routledge.

Rose, N. (1993). Government, authority and expertise in advanced liberalism. *Economy and Society, 22*(3), 283–299.

Rose, N. (1996a). Governing advanced liberal democracies. In A. Barry, T. Osborne, & N. Rose (Eds.), *Foucault and political reason* (pp. 156–188). London: University College London Press.

Rose, N. (1999). *Powers of freedom: Reframing political thought*. Cambridge: Cambridge University Press.

Rose, N., & Miller, P. (1992). Political power beyond the state: Problematics of government. *The British Journal of Sociology, 43*(2), 173–205.

Sheaff, R., Marshall, M., Rogers, A., Roland, M., Sinbald, B., & Pickard, S. (2004). Governmentality by network in English primary care. *Social Policy and Administration, 38*, 89–103.

Waring, J. J. (2005). Beyond blame: Cultural barriers to medical incident reporting. *Social Science & Medicine, 60*, 1927–1935.

Waring, J. J. (2007). Adaptive regulation or governmentality: Patient safety and the changing regulation of medicine. *Sociology of Health & Illness, 29*, 163–179.

Chapter 6
Restratification and Revalidation: United Kingdom and International Perspectives

Abstract This chapter explores the impact of the process of restratification on professional practice and medical regulation by looking at the implementation of revalidation. In doing so, this chapter acts as a necessary introduction to the exploration of the handling of fitness to practice cases in Chap. 7. Chapter 5 ended by noting that in addition to state intervention to reform medical regulation and re-establish the General Medical Council as an independent regulatory body, over the last decade, medical elites themselves have increasingly acted to strengthen peer review and appraisal mechanisms as they have sought to maintain professional self-regulatory privileges within the broader transformative context that is the emergence of the new neo-liberal governing conditions associated with the risk society. It was highlighted that the esoteric nature of medical expertise means that regardless of calls for greater consumer and inter-professional input into medical regulation, some semblance of medical control will be retained as peer appraisal remains the key mechanism by which the quality of medical work can be judged and its quality assurance assured. This chapter explores the consequences of this for the development of revalidation. This is the process by which the competence of medical practitioners will be quality assured in the United Kingdom. This chapter outlines how revalidation will operate when it is implemented sometime after late 2012. For comparative purposes, this chapter also outlines recent international trends in the quality assurance of medical practitioner's continued competence. In doing so, this chapter highlights how a move towards increased professional accountability seems to characterise medical governance frameworks internationally. This provides some empirical support for the restratification thesis. Yet this chapter also points out that it is important to note that the situation internationally is more complex than the restratification thesis allows. The need to empirically explore this point further is taken up in Chap. 8. End-of-chapter self-study tasks are provided so the reader can engage in further study in relation to chapter contents.

J.M. Chamberlain, *The Sociology of Medical Regulation: An Introduction*,
DOI 10.1007/978-94-007-4896-5_6, © Springer Science+Business Media Dordrecht 2013

Introduction

Chapter 5 ended with a clear statement of intent for subsequent chapters: the purpose of this chapter and Chap. 7 is to examine the impact of contemporary reforms in medical governance on rank-and-file practitioners and the elite regulatory bodies which oversee their training and discipline. What is more, in doing this both chapters also seek to identify how far recent developments in medical governance provide empirical support for the restratification thesis. As previously discussed, the restratification thesis reminds us that the esoteric nature of medical expertise gives practitioners wriggle room when attempts are made by third parties to survey, appraise and performance manage their work (Freidson 2001). This does not mean that doctors can fully escape the technical evaluation of their work by non-medically qualified individuals or groups, particularly when this evaluation is concerned with conducting a cost-benefit analysis or ethical valuation of said activity. Also, it should be remembered that a good proportion of medical work is somewhat readily understandable by members of the public and other professional groups, particularly within the information-rich modern age in which we live today.

Nevertheless, the esoteric nature of medical work does mean that if the state wishes to reform medical regulation to make it more open and accountable, then the proactive cooperation of medical elites is absolutely essential (Stacey 2000). In short, given the highly specialised nature of medical expertise, no matter what regulatory reforms are proposed (or finally introduced), peer appraisal will continue to play a key role in the quality assurance of medical work and the competence of individual practitioners, at least for the foreseeable future. Medical elites therefore must be co-opted into any attempt to promote regulatory reform to enhance transparency and accountability as well as reduce the threat posed by medical risk. It is this fact which hands medical elites a powerful bargaining chip when it comes to implementing change.

Yet even this advantage must be placed against the broader governing conditions of the risk society which require experts be more accountable for their activities while at the same time paying greater attention to the voice and rights of the individuals who utilise their services so they can exercise greater personal autonomy over their decision-making processes (Rose 1999). Hence, the medical profession is increasingly splintering into more pronounced elite and rank-and-file segments, based around universities, medical research communities, the royal colleges, as well as NHS bodies such as the National Centre for Clinical Excellence. Elite members of the profession are increasingly acting to subject rank-and-file practitioners to formal mechanisms of performance appraisal as they seek to respond to calls from NHS management and patient rights and consumer groups to increase the transparency and accountability of quality assurance frameworks in an effort to reduce the threat of medical risk.

Chapter 7 explores this development in relation to medical complaints and the GMC's role in the hearing of fitness to practice cases. It examines if doctors are becoming subject to greater performance management mechanisms by using statistical data pertaining to patient complaints and the outcomes of fitness to practice cases over the last decade and a half, to establish trends in GMC activity. In finding out if medical practitioners are indeed becoming subject to more punitive or rehabilitative action in relation to patient complaints using available descriptive statistical data, Chap. 7 specifically examines the continued validity of the restratification thesis. This chapter acts as an introduction to this analysis through examining another key development in the performance management of doctors in the United Kingdom: revalidation.

As was outlined in Chap. 3, revalidation involves the regular checking of a doctor's fitness to practise and is currently planned for introduction in the United Kingdom from late 2012 onwards. What this means in practice is that there currently is no hard statistical data pertaining to the operation of revalidation from which to explore its impact on medical practice, particularly in relation to the activity of rank-and-file medical practitioners, but also in the context of how medical elites undertaken revalidation processes and tasks. Therefore, we must expand our analysis if we are to begin to examine the continued legitimacy (or not) of the restratification thesis as a conceptual framework from which to explore contemporary developments in professional regulation. Here, it is important to remember that it is possible to take a step back and look at how the proposed introduction of revalidation in the United Kingdom compares with current arrangements in other countries for ensuring medical practitioners stay up to date and fit to practise in their chosen specialty. After all, if the restratification thesis is correct, then one would expect the increasing segregation of medicine into elite to rank-and-file practitioner roles and the introduction of performance surveillance and management mechanisms to be a generalisable phenomenon given that health-care systems and state governing systems worldwide are adjusting to the shifting globalised conditions of the risk society (Stacey 1992). Consequently, from looking at the introduction of revalidation through an international perspective, we should be able to explore the applicability of the restratification thesis within a broader regulatory context in order to identify key thematic trends before we turn to examine the handling of patient complaints and fitness to practice cases in Chap. 7. Let us begin by first looking at contemporary developments in the United Kingdom.

Medical Professionalism Revisited: Maintaining Competence to Practise

The introduction of revalidation in the United Kingdom was discussed in Chap. 3. One of the key themes running through this earlier discussion was the general reluctance, on behalf of elite groups within the medical profession, to address what is arguably a foundational issue for any regulatory quality assurance regime: how to ensure that a medical practitioner is competent to practise. Making sure a medical practitioner has the necessary knowledge and skills appropriate to the point in their professional career is undoubtedly a highly complicated matter. Not least of all because it requires, we accept that medical careers are not linear, people can leave and come back to work for a myriad of reasons, while the fast-paced nature of progress in medicine means educational providers must adopt a dynamic approach when defining curricula content. But it is important to begin our discussion by noting that it is only in the last two decades that the issue of how best to ensure a doctor remains competent to practise has actually begun to be addressed at all. Indeed, up until the 1980s in the United Kingdom, there were no formalised educational and regulatory systems to ensure experienced medical practitioners remained competent and fit to practise once they had achieved the status of consultant or general practitioner (Irvine 2003).

This is an important point. It is it generally held that the professionalism of a medical practitioner drives them to pursue excellence and maintain their competence. Not least of all because it is in the best interests of their patients for them to do so. In short, doctors can be trusted to stay up to date and fit to practise in their chosen specialty because they are ethically driven to do so: no good doctor wants to harm their patients let alone provide them with substandard advice and care. Now, it is certainly the case that the vast majority of medical practitioners do wholeheartedly ascribe to this fundamental ideal of what constitutes a good doctor. In doing so, many give up precious personal time outside of their already hectic work schedule in order to ensure their patients benefit from contemporary advances in medicine and that their clinical skills and medical knowledge remain second to none. Yet no matter how persuasive this idea is for both doctors and members of the public alike, it depends on two factors, namely, that a doctor does act to maintain their fitness to practise, and that they are able to keep track of current developments in their chosen specialty. Both of these factors are clearly problematic, as we shall now discuss.

Let us leave aside the complex and thorny issue of the ability of a doctor who is incompetent to know they are indeed incompetent, and furthermore, proactively seek the training and support they need to become competent. Even when we do this, it needs to be acknowledged that even a competent conscientious doctor who knows where to go to get training may from time to time fail to remain completely on top of current developments in their chosen field. Hence, there is often the need for collegiate support and organised continuing professional development activity. Yet the fact of the matter is that as the twentieth century progressed, medical elites failed to ensure the necessary educational and regulatory frameworks, and associated quality assurance mechanisms were in place to effectively deal with the rapid advances in medical knowledge and technology that were occurring. After the 1858 Medical Act, a series of subsequent Acts progressively refined the purpose and content of medical education and, in principle at least, extended the GMCs' powers in overseeing its quality. In no small part, these reforms were introduced because of the rapidly expanding nature of modern medicine and the explosion of new medical technologies and specialties which occurred during the twentieth century. So the 1950 Medical Act extended the period of university-based basic medical education from 5 to 6 years by including what was called preregistration year (which is now called foundation year one), while the 1968 Act had formally established higher specialist hospital training to be overseen by the royal colleges. In response to the growth in medical knowledge, the Act also made it clear that the purpose of basic university medical education was not to produce the finished article but rather to provide a basic grounding in medicine and prepare junior doctors for subsequent hospital-based specialist training, before they finally moved on to higher vocational training in medicine, surgery or general practice. Such changes in the focus of basic medical education, the addition of the preregistration year alongside the introduction of more formal arrangements for later specialist training, were all necessary due to the rapidly expanding nature of medical knowledge and expertise. In short, changes in the organisation of medical education occurred side-by-side biomedicine's increasing reliance upon new developing forms of medical technology. But for many, they did not go far enough, particularly in relation to continuing medical education for established practitioners.

The development of medical training rapidly expanded during the 1980s and 1990s in response to medicine's seemingly ever-expanding knowledge base. Undergraduate curricula were reformed nationally in the mid-1990s under the banner of producing *Tomorrows Doctors* (see GMC 1993). It was recognised that medical technology and expertise had progressed so far and quickly that medical schools urgently required a dramatic overhaul and needed to focus more than ever before on ensuring medical students knew how to be life-long learners and hence possessed the necessary educational skills to be able to keep themselves up to date and fit to practise (Irvine 2003). Reforms were also made to postgraduate and specialist training during this time to similarly focus on ensuring the quality of educational provision and its quality assurance. Yet it was a different story for more experienced consultants and general practitioners. In part, this was because medical elites feared that too much intervention in this regard may constrain practitioner's autonomy and clinical freedom, which were felt to already be under significant pressure from the NHS reforms discussed in Chap. 2. But the fact that the royal colleges and GMC dragged their feet on this matter worried the government. Not least of all because concerns were expressed about the ability of the GMC and royal colleges to self-police the matter. So the government commissioned an investigation into the matter. The resulting report argued for more formal arrangements for post-specialist training, called variously continuing medical education (CME) or continuing professional development (CPD) (Calman 1993). Calman followed this report with a further one on the topic of doctors continuing medical education. Here, he argued that *'the case for CME rests heavily on the concept of confidence: clinicians must command the confidence of the patients they treat; of the public as a whole; of the hospital managers to whom they are accountable for the quality of service to patients'* (Calman 1994: 6). This state of affairs made it clear to members of the profession that it was the state, not the medical profession's elite institutions, which were looking seriously at doctors continuing competence to practise. A direct state attack on the principle of medical self-regulation was felt to be under way. In an editorial in the *British Medical Journal* (BMJ), Richard Smith (1993: 974) held that *'the government is sidelining the GMC and with it the self-regulation the profession has enjoyed since 1858'.*

However, despite a global recognition amongst medicine's elite institutions of the need to act little had actually been done, this raises the question: 'Just what was going on?' Although the GMC and the royal colleges recognised the need to act on the issue of doctors continued competence to practise, the *'royal colleges were most interested in making sure the GMC did not stray seriously into their territory, into specialist training or CPD, or indeed act decisively on its statutory duty to co-ordinate all stages of medical education'* (Irvine 2003: 98). The colleges had introduced their own more formal arrangements for CPD. They worked with the medical professions key trade union, the British Medical Association, alongside the state, to establish mechanisms whereby sanctions were introduced for doctors who failed to gain CPD 'points' for completing college courses, such as exclusion from merit awards and the supervision of junior doctor training posts (Boulay 2000). Yet these lacked the key sanction possessed solely by the GMC: removal from the medical register for non-completion of CPD. In summary, when the medical profession elites recognised something needed to be done, a tendency towards institutional inertia and the protection of doctors from external surveillance and control both remained. To some extent, this was to be expected. After all, the medical profession's exclusive cognitive identity advocates a form of mutual protectionism, which frowns on breaking ranks and whistle blowing (Stacey 1992, 2000). As we explored in Chap. 3, this was symptomatic of the exclusive closed shop mentality of the medical club of the time.

It has to be acknowledged that there was a degree of pressure from within the profession, demanding that the GMC become more proactive and take up the challenge of underperforming doctors, particularly from general practitioners, but also from powerful insider commentators such as Richard Smith, who was editor of the BMJ (i.e. Smith 1992). These reformers felt the GMC was too far removed from the needs of the profession. They wanted it to provide definitive leadership to its rank-and-file members by forging a more open and accountable relationship with the public. One of these reformers, who was heavily influenced by the sociologist and GMC lay member Margaret Stacey, as well as the medical sociologist Margot Jeffreys, was a general practitioner called Dr Donald Irvine (now Sir Donald Irvine). He would be the first leader of the GMC to be a general practitioner since its foundation 137 years previously. Irvine (2003:11) noted: *'In 1995, I stood for election as President of the GMC, on a programme of reform both of professionalism in medicine and the GMC itself. There were members within the GMC, both medical and lay, who believed that such reform of the GMC had to be carried out swiftly. Otherwise public confidence in the medical profession, and in particular in the system of professional self-regulation, for which the GMC was primarily responsible, could not be sustained'*. Irvine oversaw the GMC during a particularly turbulent time in its history: the Shipman case was to no small degree to come to define his period in office. Indeed, although it is arguable that Irvine was not able to push through the reforms he wanted due to opposition of the royal colleges in particular, nevertheless, he was instrumental in making sure the issue of revalidation – that is, that a doctor must periodically prove their competence and fitness to practise under the watchful gaze of training and regulatory elites – remained at the forefront of plans to reform medical regulation and modernise the GMC.

Revalidation in the United Kingdom: A Developmental Process

Chapter 2 outlined the key developments which occurred in relation to revalidation in light of the Shipman case and the subsequent review conducted by Dame Janet Smith (Smith 2005). As was noted, Smith (2005) was extremely critical of annual appraisal and the GMC's proposals for revalidation. As a result of her criticism, it was proposed that revalidation would be compulsory for doctors to stay on the medical register (The Secretary of State for Health 2007). It was felt that the revalidation process should involve a mixture of clinical audit, direct observation, simulated tests, knowledge tests, patient feedback and continuing professional development activities. It was proposed that revalidation should be made up of two elements – relicensing and recertification – and incorporate NHS appraisal within it. Relicensing, it was argued, would make current NHS appraisal arrangements more rigorous, with greater direct testing of a doctor's competence in regard to key day-to-day clinical tasks being overseen by the royal colleges. However, although revalidation was originally planned for introduction in 2010, the development and piloting process has taken somewhat longer than expected. A consultation process currently under way at the time of writing, which involves extensive public, managerial and professional involvement, as well as a series of what have been termed pathfinder projects to test out and refine the revalidation process, is unlikely to be completed till mid-2012 at the earliest. However, at this point in time, that is, prior to wholesale national implementation, it is possible to outline what its likely form is to be, as well as relate the trend towards greater peer surveillance and control in the United Kingdom to recent developments in medical regulation within the international context, in order to ascertain if there indeed is a more global trend towards increased professional accountability within the implementation of medical governance and quality assurance frameworks. Let us begin this process by examining contemporary developments in revalidation.

The introduction of revalidation in the United Kingdom is overseen by the Revalidation Programme Board (RPB). Membership of this board includes the academy of medical royal colleges, the GMC, the British Medical Association, National Health Service employer representatives as well as the four UK health departments: the England Department of Health, the Scottish Government Health and Wellbeing Directorate, the Welsh Assembly Government and the Social Services and Public Safety Northern Ireland Executive. The board is responsible for ensuring the timeline and milestones for the implementation of revalidation are adhered to as it is designed, planned and rolled out nationally across England, Wales, Scotland and Northern Ireland. When first proposed, as already noted, it was expected that revalidation would consist of two processes: relicensing and recertification. Relicensing – sometimes called relicensure – was to be based around the generic standards of practice set by the GMC and a much revised and more formalised version of NHS annual appraisal, overseen by a medical director. The revised annual appraisal process involves multidisciplinary feedback from medical colleagues, other health- and social-care professionals, administrative and support staff as well as patients (Cato 2008). For this reason, this is often termed 360-degree feedback.

Recertification was to occur every 5 years, be overseen by the royal colleges, and involve a comprehensive assessment of a medical practitioner's competence. Although not fully defined when first proposed, it was expected that recertification would involve some form of hands-on testing of a doctor's clinical skills and knowledge. Taken together, this initial design for the revalidation process would have provided a regular annual check-up alongside a 5-yearly rigorous examination of a doctor's clinical competence and fitness to practise. However, during the exploratory phase of the development of revalidation as the early pathfinder projects reported on the feasibility of the process, the decision was taken to simplify the process and build it solely around a doctor's annual appraisal. In part, this was to do with costs and the difficulties involved in recertifying a large number of practitioners. Yet it was also felt that it was important to not just embed revalidation within existing employment and practice systems, but rather to also reinforce the developmental and formative nature of revalidation. In linking revalidation firmly within the context of a rigorous annual appraisal of a medical practitioner's overall clinical performance within the work setting, it was hoped that it would become a meaningful exercise in ongoing professional development. As the GMC stated, *'revalidation will neither be a box ticking exercise nor a punitive process. What it will be is relevant to our day to day medical practice and built upon systems that should already exist in the workplace to support high quality care. The GMC and other organizations are determined to make sure that it will neither be burdensome nor hamper in any way in fulfilling our main duty, caring for patients'* (GMC 2010a:1).

It is undoubtedly the case that the introduction of revalidation is in principle a threat to medical autonomy in the sense that it requires doctors and their regulatory institutions be more open, inclusive and accountable for their practices. In short, the introduction of revalidation can be seen to be bound up with the contemporary shift towards increasing governmental control over the operation of health systems and the regulation of the medical profession. It is certainly the case that medical elites, specifically the GMC and the royal colleges, were dragging their feet on the matter of how best to ensure practitioners remain fit to practise in their chosen specialty, at least until the mid-1990s. While the idea of revalidation was not initially welcomed by the medical profession en masse, indeed, it was arguably only accepted as being necessary by certain sections of it when it became clear that there was no other option but to accept it (Irvine 2003). Yet what we have seen since revalidation became enshrined in governmental legislation via the 2008 Health and Social Care Act is an increasing medical co-option of its form, content and process, predominately by medical elites operating in the various royal colleges, as they have sought to ensure that an acceptable balance is struck between the care and control aspects of revalidation. Indeed, this is perhaps most recognisable in the shifting emphasis of the form and focus revalidation.

Originally, when first legislated for, the risk surveillance and control element of revalidation was emphasised via a rhetorical focus on protecting members of public from doctors who are not fit to practise. Hence, the recertification element was proposed as being a 5-yearly rigorous test of an individual medical practitioner's competence to practise. Indeed, the president of the GMC at the time, Sir Graham

Catto, often referred to it in terms somewhat similar to the MOT road test cars have to pass in order to be considered road worthy in the United Kingdom (Catto 2006). But over time, as the practicalities of revalidation have been worked out, this emphasis has shifted towards making it into a more developmental 5-yearly cycle. So all medical practitioners will now experience a more rigorous form of annual appraisal under the watchful gaze of their responsible officer, who will typically be their medical director. Indeed, while the process will be overseen by the GMC in day-to-day practice, revalidation will be quality assured by the royal colleges working in tandem with the responsible officer.

What will happen then is that all licenced medical practitioners will be linked to a responsible officer who will make a recommendation to the GMC about their fitness to practise based on the outcome of their annual appraisals every 5 years but on the basis of progress achieved over the 5-year period. Licenced practitioners will have to maintain a personal portfolio of supportive evidence of the fitness to practise, and in no small part the material therein will be based around the GMCs guidance on the duties of a doctor as well as NHS clinical governance systems (GMC 2010b). It is expected that key elements of the portfolio will include responses to patient feedback and formal complaints, medical and clinical audits, educational activities, clinical team 360-degree feedback, as well as research activity outcomes. The focus, therefore, is on making sure a practitioner is responding to issues and problems positively in a developmental manner while at the same time demonstrating an appropriate level of technical competence given the stage of their career.

Restratification, Revalidation and the Health Select Committee

In essence, what we have is a revalidation process which is medically controlled in the sense that although it is open to third-party input (in the form of NHS management, other health- and social-care professionals and patients), as well as is subject to independent oversight by the GMC, nevertheless, it fundamentally relies on the mechanism of peer appraisal and assessment to make it work. Indeed, it is the royal colleges who have provided the specialist performance frameworks and standards from which to make revalidation work (Starke and Brownbridge 2010). What is more, the focus on embedding revalidation within a 5-yearly cycle reflects a growing emphasis on ensuring it is a constructive and developmental exercise that should protect patients from a small number of underperforming doctors who are not fit to practise while at the same time providing reassurance that the majority of doctors are keeping up to date, reflecting on the quality of care they provide, as well as ensuring they develop as practitioners as their career progresses. Here, revalidation needs to take into account the fact that some doctors work on a locum basis as well as take career breaks due to illness or career choice, and indeed, the current developmental and piloting process has revolved around ensuring that the system is flexible enough to cope with the diverse career pathways practitioners can follow (NHS Revalidation Support Team 2009).

In summary, as it has developed, revalidation is no longer so much concerned with identifying ‘bad apples’ (although it should do this) as with affirming good practice and ensuring it is recognised locally by employers and work colleagues. To be sure, as long as doctors are collecting evidence of their performance and professional development consistently, then the revalidation process will be very much ‘light touch’, at least in terms of its level of direct interference with their practice. Nevertheless, one qualitative study conducted as part of a revalidation pathfinder pilot did find that although most of the general practitioners interviewed felt the collection of work-based supportive material operating in a 5-year cycle for revalidation was feasible, work related and access constraints meant that some concerns were voiced regarding the appropriateness of patient feedback and some of the newer types of supporting information, that is, learning credits (Charlton et al. 2011). Such essentially technical matters should be resolved as the piloting stage is completed. But it should be noted that other practitioners have expressed concern over the potential for revalidation to be an overly bureaucratic and standardised process concerned with certainty and safety to the determent of promoting medical professionalism (Lynch 2010).

It is certainly the case that the introduction of revalidation has caused fear and anxiety amongst some quarters of the profession. This in no small part is why the GMC and royal colleges are emphasising its developmental, cyclical, nature. Yet to some extent, this state of affairs is to be expected as there is a somewhat natural tension between bureaucratic managerial systems of surveillance and control which seek to standardise working practices to make them measurable and predictable, and professions such as medicine which emphasise practitioner autonomy in the form of freedom of judgement based around the possession of specialist knowledge and expertise alongside a recognition of the inherently messy nature of the real world of professional practice. Furthermore, as Chaps. 4 and 5 discussed, the sociological study of the professions is replete with reminders concerning the ability of members of the medical profession to co-opt managerial imperatives for their own ends. Nevertheless, what is interesting about the current situation is that whereas previously this tension between managerialism and professionalism possessed clear in-group and out-group elements, that is, medicine and NHS management, over the last decade, fractures have opened up within the medical profession as elite groups have sought to instigate change by placing rank-and-file practitioners under greater peer surveillance via performance appraisal mechanisms, of which revalidation is but the latest example (Chamberlain 2009).

The current unfolding situation appears to reinforce the legitimacy of the restratification thesis: rank-and-file practitioners are indeed coming under increased peer surveillance and performance appraisal and in doing so are becoming more accountable for their practices. Yet, at the same time, it is also important to note that reforms to the structure and organisation of the GMC, alongside the introduction of revalidation, reinforce the more proactive role being played by the state in constraining medical autonomy, both in the form of the principle of professional self-regulation and in terms of clinicians clinical freedom at the patient bedside. Perhaps the key outcome of state intervention is then that there appears to be a trend towards, on one hand, reforming the GMC so it is an independent regulatory body and much less dominated by representatives from medical elites as it has historically been, while on the other hand, there also seems to be a shift towards increasingly relying on medical elites to deliver regulatory reform and this has led to a real increase in their autonomy and control over medical training and regulation as they increasingly act to place the work-based activities of rank-and-file practitioners under regular scrutiny. Here, it must be remembered that while medical elites in the form of the royal colleges form a key part of this enterprise, so do their close cousins working in what we can call, for want of a better term, the guideline industry. In short, the regulatory picture is more fractured and complicated than ever before, and it appears that the nature of medical autonomy is being transformed. Hence, while the restratification thesis goes some way to explain what is happening, it is also important to remember that it is the state which has acted to open up medical regulation in response to a broader need to reform how professionalism operates as a regulatory strategy under the risk saturated conditions of contemporary society.

This is a point which we will return to later in this chapter as well as in Chaps. 7 and 8. For the moment, it is important to end our discussion of the introduction of revalidation in the United Kingdom by noting that a key outcome of this shift in the governance of medical regulation is that it has become open to even more disagreement and flux. Not least of all because if medical regulation is truly to be open and accountable, then the voices of both patients and other health- and social-care professionals must be listened to and included within the regulatory reform agenda. What this means in practice is that one is more likely to encounter disagreements and compromises then previously was the case. No longer is it a matter of disparate elements of the medical profession arguing with each other to effect (or resist) change, particularly when the principle of medical self-regulation is directly called into question, as did by and large happen during the 1970s, 1980s and even in the 1990s. Patient groups, NHS employers and managers, alongside other health- and social-care professionals, are all now recognised members of the conversation possessing a direct stake in shaping its future.

We can perhaps see this change most clearly in the current governmental response to the GMC consultation exercise in relation to the exact nature of the licence to practise and revalidation regulations to be put before parliament in 2012 (GMC 2010b). The regulations requiring the implementation of revalidation under the 2008 Health and Social Care Act came into effect in October 2009. However, this just started the ball rolling in relation to the development of revalidation. The subsequent consultation exercise had to account for many important issues, including the conditions under which an individual whose licence to practise has been withdrawn can regain their licence. During the consultation process in June 2010, the secretary of state for health decided to extend the piloting period to develop a clearer understanding of the costs and practicalities surrounding the implementation of revalidation. As part of this process, the House of Commons select committee heard evidence from the GMC, members of the public, NHS and private practice employers, hospital management as well as other interested parties, concerning the nature of the revalidation process and the proposals for its oversight by the GMC (House of Commons Health Committee 2011a).

The committee made several key recommendations in relation to the changing role of the GMC and revalidation. First, it was noted that currently the GMC is accountable to parliament via the Privy Council. However, revalidation has brought to the foreground the question of the effectiveness of this accountability, not least of all because by and large the GMC has been left alone by parliament to run its own affairs. Hence, it was decided that the health select committee would exercise the accountability function held by the Privy Council under the 1858 Medical Act and require the GMC every year to provide oral evidence – currently, it only has to provide an annual written report of activity – so it can be actively questioned on operational matters by committee members. In effect, this will serve to add another state-actioned governmental surveillance and control regulatory layer to existing medical governance frameworks.

Second, the committee reiterated that revalidation must begin from late 2012. In no small part, this is because the government does not want there to be more delays in the implementation of revalidation as it is seen to be essential to ensuring the effective risk management of underperforming doctors. This can perhaps be most clearly seen in the committee's third recommendation. This reiterates that although the health committee agrees with the GMC that revalidation is an opportunity for doctors to demonstrate their commitment to practise improvement, nevertheless it is important that the public be protected from dangerous doctors: poor performance which places members of the public at risk should be identified and tackled at an early stage by responsible officers, the GMC and NHS employers. Hence, where needed, the GMC must be prepared to suspend a doctor and investigate their fitness to practise. Indeed, the fourth and fifth recommendations of the health committee are particularly concerned with this very issue. The committee argues that revalidation is being introduced to sustain public confidence in the medical profession, and an important part of this lies in ensuring that both patients' and doctors' rights are respected and safeguarded; however, it is also noted that the aim of revalidation is to protect patients and their needs must take precedence. In particular, the committee is worried about the use of the term remediation by the GMC in relation to the investigation of issues, as this implies that doctor's needs may come before patients and in essence prejudge the response when concerns are raised. Indeed, the committee argues that dealing with doctors whose conduct or performance is of concern is primarily a local clinical governance matter and if revalidation is to be effective, then action needs to be taken long before a matter becomes serious enough to warrant the GMC undertaking formal fitness to practice procedures. In practice, what this means is that local systems must operate rigorously and effectively, and the health committee's sixth recommendation is that the responsible officer who is overseeing the process must be provided with more specific guidance on the handling of cases where concerns are raised in order to ensure patient need is prioritised.

The health select committee's seventh recommendation pertains to the strengthening of annual appraisal as a key part of the revalidation process. One of the main findings of the pathfinder pilot projects is that practice varies across the country in regard to appraisal. The committee states that the GMC needs to ensure that appraisal is a robust and consistent process which is rigorously embedded within clinical governance frameworks. Part of this, it is argued, involves clarifying the role of the responsible officer to ensure the potential for role conflict is minimised. In particular, the committee wants guidance to be issued and safeguards out in place to ensure not just patients are protected but that doctors are also protected when a conflict of interest may have influenced the decision of a responsible officer. Finally, the committee recommends that the GMC ensures that the revalidation process takes into account the need to ensure doctors who follow a varied career pattern as well as that the revalidation process ensures a doctor's ability to communicate effectively with patients. Taken together what the recommendations of the health select committee reinforce, the regulatory state is willing to take steps to ensure that the GMC, as an independent regulator, is acting in an open and accountable manner and placing the safety of patients at the centre of its activities.

The response of the GMC to the health committee recommendations demonstrates that there has been a fundamental shift in the working culture of the GMC over the last decade. The GMC stresses its commitment to ensuring patients are protected from poorly performing doctors and that it will act to ensure a robust revalidation system is in place by the end of 2012 which protects patients as well as rewards good practice (House of Commons Health Committee 2011b). The GMC stresses the need to work in partnership with the NHS, patient groups and other health- and social-care professionals, as it seeks to ensure doctor's competence, alongside their ability to produce appropriate information using NHS information systems, are all risk managed in such a way that revalidation can be implemented at a national level as smoothly as possible. In conclusion, although it is not possible to state with certainty at this time if revalidation will achieve its aims, indeed it will be necessary to collect empirical outcome data for several years after its implementation before any conclusions can be made; nevertheless, it does seem that the basic thesis of the restratification perspective holds true: medical elites are indeed subjecting rank-and-file practitioners to greater peer surveillance and control mechanisms. Yet, as already noted, broader changes also seem to be afoot: the state has acted to change the governing conditions under which medical autonomy must operate. It seems that the restratification process is not preserving medical autonomy in its traditional form but is rather transforming it into a new form.

Bearing this in mind in the rest of this chapter, we will turn to explore recent trends in how several other countries around the world act to ensure a medical practitioner stays up to date and fit to practise in their chosen specialty. But before we do this, it is important to note that each country discussed possesses specialist associations which typically oversee arrangements for ensuring doctors keep up to date and fit to practise through engaging in continuing medical education, or as it is also referred to, continuing professional development (Peck et al. 2000). Typically, elites in the form of medical associations who oversee the provision of continuing professional development activity tend to promote it as if it constitutes good medical practice and set out key competency frameworks and accredit training programmes in relation to these to ensure practitioners can maintain their fitness to practise. Yet it has to be acknowledged that considerable variation exists in how different countries go about organising, delivering and quality assuring training programme provision. Furthermore, it is not the purpose of this chapter to explore such matters in detail and outline specific differences between countries worldwide as it does so. Not least of all because such an enterprise is beyond the scope and purpose of this book. Rather, the focus is on broadly identifying if current developments provide support for the restratification thesis in terms of there being a trend or not towards more formal peer appraisal mechanisms and increased professional accountability to the public via increased external surveillance of regulatory and training activity. Let us begin this exercise by examining medical regulation in Australia.

Australia

Australia follows a federal health-care model in which each of the states which make up the country plays a key role in funding and delivering health care via the Medicare system. Doctors are remunerated via a fee for service system and are affiliated to specialty medical colleges. Similar to the United Kingdom, Australia saw from the mid-1980s onwards a growth in patient complaints, in relation to both the health-care system and the medical profession, alongside the development of patient rights movements. Throughout the 1980s and into the 1990s, the government has focused on reducing adverse events within the health-care system and reforming medical regulation as a result of a series of high-profile medical malpractice cases (Tito 1996). This reforming process has continued into the twenty-first century, with the establishment of the Australian Council for Safety and Quality in Health Care, which is tasked with developing guidelines and standards for care. This governmental body works with the state medical boards which oversee the standard of medical training and practice in tandem with various local level health-care agencies. A central medical database of registered practitioners is overseen by the Australian Medical Council. Medical boards oversee matters of discipline and will temporarily suspend or strike off a doctor under the egis of the 1992 Medical Practice Act, if need be.

As a result of state pressure for change and patient right group lobbying, in 2000, all medical boards made it a requirement for doctors to make an annual declaration of their fitness to practise. This declaration requires a doctor to state their qualifications and experience, give their health status, provide details regarding their current employment, provide information regarding any criminal convictions, professional warnings or misconduct as well as detail their completion of specialty college continuing medical education activity. The role of the specialty colleges is then vitally important in overseeing the continued fitness to practise of doctors. They work in tandem with the health-care bodies responsible for addressing complaints. While since the beginning of the new millennium there has been a growing managerial involvement in monitoring health-care work as well as how the medical boards monitor and quality assure individual doctors. Indeed, it is clear that similar to the United Kingdom, medical work in Australia since the 1980s has gradually become subject to managerial performance appraisal measures, while patient rights groups have successfully lobbied for non-medical and intra-professional membership on medical boards as well as the annual declaration monitoring process (Lupton 2011).

Canada

Canada has a federally organised model similar to that of Australia. Health care is funded via federal government and regional tax revenues, but there is a burgeoning personal health plan industry. Unlike the United Kingdom but like the Australian system, doctors are paid on a fee for service basis, both in the hospital and community settings (Detsky and Haylor 2003). Medical councils oversee regional areas from a national perspective, with a Federation of Medical Licensing Authorities of Canada dealing with matters of registration and discipline nationally. However, a system of royal colleges and medical associations is in place, running across the medical specialties, and these bodies variously bear the responsibility for ensuring a doctor is fit to practise by periodically completing a peer-review process under the Regulated Health Care Professions Act 1991.The colleges and associations run a Peer Assessment Programme, with doctors being periodically subject to a review of the fitness to practise as a result of age (if over 70 a doctor has a mandatory retest of competence every 5 years), self-request or significant peer concern (usually via medical error in the hospital or community setting). Health-care management and patients can also make complaints which sometimes will lead to a doctor becoming subject to a peer test of their fitness to practise. Some colleges and associations randomly select which members peer test each year; others adopt a 5-yearly (or less) cyclic process (Gerace 2003). Peer testing usually involves a mixture of practice observation, feedback from patients and peers, as well as knowledge and clinical skills tests. All peer reviews are completed on the basis of identifying the potential for risk to patients. The reviews are very thorough and are seen as expensive and time consuming by some, while some doctors report that they can be quite intimidating affairs (Gerace 2003).

Since the beginning of the millennium, there has been a shift towards increasing managerial control of health-care provision in Canada to contain costs and risk. There has also been a concurrent increase in the number of doctors becoming subject to peer testing as a result of hospital management expressing concern over their practice. Furthermore, as with other countries, there have been calls to make medical regulation more open and transparent, which has led to an increase over the last decade in lay membership on peer-review and disciplinary hearing panels. At the same time, increasing emphasis has been placed on practitioners learning how to maintain a positive doctor-patient relationship as well as improve their communication skills. It has been argued that peer testing of a doctor has become more frequent as well as more likely to reveal a problem which requires some form of remedial educational training programme (Detsky and Haylor 2003).

Finland

In contrast to Canada and Australia, health-care provision in Finland is paid for solely by an obligatory state-managed sickness insurance programme. Although private medical care does exist, it forms a very small minority of health-care provision. The Finish government Ministry for Social Affairs and Health is responsible for policy, resource allocation and setting practice guidelines as well as overseeing quality standards. The Ministry's executive authority, The National Authority for Medico-Legal Affairs, oversees medical regulation by settings standards and quality assuring the work of health-care practitioners. It works in tandem with the National Public Health Institute which is responsible for providing both professionals and citizens with health information and evidence. Although the state is clearly heavily involved in defining health-care provision and its quality assurance, the Finish Medical Association acts in a self-regulatory manner, overseeing doctors training, registration and licensing under provisions of the Health Care Professionals Act 1994. In many ways then, the organisation of the state-profession relationship is strongly similar to that of the United Kingdom, with specialist societies acting as self-regulating provisional bodies similar to the royal colleges in the United Kingdom and feeding into the operation of the Finish Medical Association. Concerns were expressed during the 1990s about the lack of formal training for doctors at a continuing medical education level, and the Ministry for Social Affairs and Health stated in 2004 that all doctors should engage in around 10 days of such activity a year to ensure they remain fit to practise.

The specialist societies have the role of improving training provision and work in tandem with the National Authority for Medico-Legal Affairs in dealing with complaints against doctors or events which lead to their fitness to practise being questioned. But performance reviews, as they are called, are the responsibility of the National Authority for Medico-Legal Affairs, not the specialist societies. Such reviews can be triggered by a complaint from a patient, a colleague or local employer. There is no ongoing annual or cyclical test of a doctor's fitness to practise unless a problem occurs. In conclusion, it is the National Authority for Medico-Legal Affairs which possesses responsibility for a practitioners licensing and discipline, not a self-regulatory medical professional body, and Finland can consequently be said to epitomise a state-control model of professional regulation (Ministry of Social Affairs and Health 2004). However, it should be noted that similar to other countries, Finland has seen a rise in patient complaints against doctors in the last 20 years, and the state has had to act to engender reform in the provision and quality assurance of continuing medical education as it seeks to manage the costs and risks associated with health-care provision.

Netherlands

The Netherlands health-care system operates via a mixture of public and private health-care insurance, with higher income individuals tending to have private insurance. In many ways, the Dutch health-care system sits between the predominately state-funded health-care systems of countries such as the United Kingdom and Finland and those which rely on private fiancé like the United States. Yet like Finland, the Dutch government plays a significant role in the setting of health-care standards and quality assurance processes via the 1996 Care Institutions Quality Act (Lombarts 2003). The Ministry and Health and the Health Inspectorate regulate the quality of health care and establish performance indicators for good practice. Yet the principle of self-regulation dominates the system of medical regulation as it is recognised that professionals must be free to make clinical decisions. Hence, medical associations control the development of clinical practice guidelines and monitor individual doctor's continuing medical education activity. In essence, the Dutch system could be said to be much like the United Kingdom in that it involves a careful balancing act between medical autonomy and governmental control. At the heart of this balancing act lies the Royal Dutch Medical Association, which is the medical dominated body responsible for the training, registration and recertification of specialists, while the Central Colleges of Specialists defines specialty training programmes. Since 2000, the Central Colleges have required doctors to practise their specialty for a minimum number of hours per week and complete at least 40 h of accredited continuing medical education activity per year to stay on the medical register. They also undertake peer-review visits to hospitals and general practices to review doctors and their workplace practices. Such visits are undertaken on a 5-year cycle. Doctors can also have their performance reviewed at any point as a result of complaints from patients and hospital management. As with other countries, patient complaints and the number of peer-review cases undertaken annually as a result of medical error have risen in recent years with the result that there has been a slight shift towards more bureaucratic managerial and lay involvement in regulatory processes; however, the Dutch regulatory system still retains a large degree of medical control and autonomy (Swinkels 1999).

United States

In contrast to the states discussed so far, in the United States, health care is predominately privately funded through health insurance corporations and employer-based insurance schemes. However, publicly funded health insurance operates through the Medicare and Medicaid schemes. Medical practitioners are paid on a fee for service basis. It is because of this state of affairs that health providers and insurance schemes can be said to play a key role in medical regulation as they usually require doctors be properly credentialed and certified through medically dominated State Board specialty training and continuing professional development programmes. Although federal state-level medical boards are key to the training and regulation of doctors as it is these bodies which issue licences to practise, the core national-level body for ensuring standards is the National Board of Medical Examiners. This sets the examination standards from which individual state medical boards license doctors to practise. The key task of the National Board of Medical Examiners is to oversee the Medical Licensing Examination, which all doctors must pass during their residency (this training period is equivalent to postgraduate medical training in the United Kingdom). It also oversees a National Practitioner Data Bank which is similar to the medical register of approved practitioners found in other countries, including the United Kingdom. In the United States, all doctors must go through a periodic retesting of their competence to practise, which is known as recertification. Certification occurs as a result of the successful completion of initial training and residency requirements. Recertification periodically occurs as a result of the subsequent successful completion of what since 1998 has been called a Maintenance of Certification Programme (Pugh 2003). Recertification through this programme consists of a mixture of exams to test knowledge, simulated tests of clinical skills and communication skills, peer observation of day-to-day practice as well as patient feedback. Over the last two decades, increasing concerns over medical risk and cost, in no small part as a result of medical error due to the exponential growth in medical knowledge and technology, has led to Maintenance of Certification Programmes becoming more rigorous in their testing of doctors (Duffey and Zipes 2004). Recertification typically occurs between 2 and 5 years, depending upon which state medical board and medical specialty a doctor works in. It is generally held that there is a trend in the United States towards the increasing standardisation of recertification programmes at a state Medical Board level (Cain et al. 2005). Furthermore, it appears that pressure from patient complaints, health insurers and employers is driving this trend. Indeed, in line with the restratification thesis, it has been said that the medical profession is increasingly splitting into elite and rank-and-file roles, with rank-and-file doctors becoming increasingly subject to more formalised peer appraisal processes (i.e. see Freidson 2001), which is perhaps to be expected as the restratification thesis was first formulated in the context of contemporary developments in the Anglo-American regulatory context (Lupton 2011).

Critical Reflections on the Restratification Thesis

The preceding overview of medical regulation arrangements and trends in the governance of medicine in several different nation-states highlights some pertinent issues for the discussion in relation to the validity of the restratification thesis. First, it does seem that over the last two or three decades, there has been growing calls from a range of diverse stakeholders – including patients, health insurers and employers – for doctors to become subject to at the very least more formalised continuing education processes in an effort to deal with the problem of medical risk and error, with the result that since the 1990s efforts have been made by medical elites, sometimes as a result of direct state legislation, to enhance processes and practices already in place to ensure a practitioner's continued fitness to practise. Most countries worldwide now seem to operate some form of revalidation process, although the nature and frequency of such arrangements varies somewhat from country to country (Kelly 2010). The drivers for this reform are diverse, but invariably include rising health-care costs and an increasing perception on behalf of the public and governing elites that doctors cannot be left alone to manage their own affairs free from outside interference, in no small part as a result of high-profile cases pertaining to medical negligence and error (Davies 2007). However, this brings us the second issue, namely, that it is quite clear that outside groups, such as state-sponsored monitoring bodies, health-care managers, other health- and social-care professional grouping and patient rights groups, together play different and varying roles within the field of medical regulation internationally. Indeed, there is no one set way in which medical regulation operates, with most countries operating somewhere on the spectrum between the extremes of complete self-regulation and complete state regulation. To be sure, and perhaps most importantly, that is, at least in the context of the restratification thesis, although all doctors internationally possess self-regulatory institutions and professional associations which are involved in some way or other in the setting standards and the quality assurance of medical work and medical training activity, the level of direct state control in medical governance arrangements varies considerably internationally, that is, compare the United Kingdom with Finland.

It certainly is true that in the United Kingdom, occupations traditionally categorised as professions, such as medicine and law, have possessed a significant degree of occupational control over regulatory arrangements, along with a greater degree of opportunity for self-employed status like their American counterparts, than their mainly directly state-employed continental cousins. Furthermore, in European countries, such as Germany and France, state bureaucracies have traditionally controlled arrangements relating to examination, licensing, standard setting and disciplinary procedures. All of which have historically been controlled by independent professional associations in the United Kingdom. Yet, here it should be noted that medical practitioners have possessed considerable privileges on the continent despite the more direct role played by the state in managing their affairs. Indeed, they have been able to secure high levels of job security and income as well as a large degree of autonomy in their work, similar to their Anglo-American counterparts (Burrage and Torstendahl 1990).

It is not the aim of this book to undertake a comparative and historical analysis of the arrangements surrounding the organisation of occupational groups regarded as professions across nation-states. But one thing is clear, a central feature of the restratification thesis is that the medical profession's specialist knowledge base leads to peer review forming an essential and core feature of any regulatory arrangement. This seems to hold true internationally, regardless of the level of state involvement professional associations are heavily involved in standard setting, monitoring and performance appraisal arrangements. Furthermore, even a cursory review of arrangements internationally reveals that the conditions at the very least appear ripe for medical elites to subject rank-and-file practitioners to greater peer surveillance and performance management mechanisms. It does indeed appear to be the case that since the 1990s, medical elites internationally have acted to strengthen peer-review mechanisms via introducing either more formalised continuing medical education activity arrangements or introducing a formal revalidation-like process

(Mekur et al. 2010). But it is equally clear that the applicability of the restratification thesis needs to be explored, and its assumptions perhaps revised, in light of any review of international trends.

This leads us to our third and final point, which is that quite clearly, if the restratification thesis is to be developed further, then a research agenda needs to be established for the exploration of current international trends in the monitoring of medical practitioners continuing competence to practise. Such a research agenda will need to focus on the historical and comparative development of medical regulatory arrangements as well as how the relationship between the state, the public and the medical profession has developed over the last several decades. It will also need to focus on the relationship between rank-and-file practitioners and elite groupings of the medical profession in order to identify if there is a trend towards the increased formal monitoring of professional practice. One would therefore expect empirical work to be undertaken with patients, rank-and-file doctors, representatives from elite bodies, as well as quite possible hospital managers and other interested parties. The goal being to identify if the assumptions and claims of the restratification thesis are correct or of another explanation for what is happening would be more appropriate. Research outcomes would need to be compared internationally across nation-states to identify global trends: it could well be the case that the restratification thesis is more applicable in some countries and not others. This need to establish a research agenda from which to explore current trends in medical regulation, along with their consequences for both doctors and patients, will be touched on again in Chap. 8. For the moment, it is necessary to return to the introduction of revalidation in the United Kingdom and conclude this chapter with a brief summary of the key points made so far.

Conclusion

Chapter 5 has outlined the implementation of revalidation in the United Kingdom. In doing so, this chapter highlighted the initial reluctance of medical elites to reform the essentially voluntary arrangements in place for helping a doctor maintain their fitness to practise. Indeed, it was highlighted that the royal colleges in particular initially acted to protect medical autonomy by introducing continuing medical education programmes only when it became clear that they had no choice but to respond to state pressure, as well as by and large acted in a similar manner when the idea of formalising a periodic test of a practitioner's continued fitness to practise was first mentioned in the 1990s. Indeed, it is unfortunately the case that elite groups within the medical profession have arguably not responded to the changing broader sociocultural context and public expectations of modern medical practice in a manner that they perhaps should. Instead of taking the lead and driving through necessary reform to medical regulation from an internally generated programme of change to reform how the medical profession manages its affairs, it seems to be the case that medical elites have only responded when it has been necessary to do so as a result of direct state intervention into medical governance (Irvine 2003). This is not to say that the need for change has not been recognised from within the profession or that some medical elites have not acted to try and bring around regulatory reform. Yet, as Chap. 3 of this book outlined in some detail, the fact of the matter is that it is only when the state has acted through enacting legislation that the profession has responded and sought to reform medical regulation. This is particularly the case with regard to revalidation. It was only as a result of state legislation – the 2008 Health and Social Care Act – that the fitness to practise of doctors in the United Kingdom will be periodically quality assured in some form. But even here, as this chapter outlined in some detail, medical elites have acted to put pressure on the state to shift the regulatory emphasis of revalidation away from the summative testing of practitioners and towards the more formative affirming of good practice while also recognising areas for development.

The response of medical elites towards the introduction of revalidation in the United Kingdom to some extent provides evidence for the restratification thesis. In that, it is clear that the medical profession still plays a powerful role in deciding how regulatory processes and their quality assurance will be undertaken. It is also clear that medical elites are subjecting rank-and-file practitioners to performance appraisal, even if it is formative and developmental by nature, rather than summative and punitive. Yet, as this chapter noted it is also apparent, when the international context is looked at, that although it is possible to identify broadly similar trends in some countries, overall, the regulatory situation is far too complex to be clear cut. In short, the restratification thesis may well need some refinement. Hence, a sustained programme of research is necessary. Importantly, although there is a clear need to look at the issue of how a doctor's fitness to practise is monitored and quality assured by both medical elites and the regulatory state, it is just as necessary to look at the management of disciplinary processes and hearings. Not least of all because in doing so, we should be able to identify the impact of regulatory reform on both patients and doctors. This leads us to the next chapter of this book, which explores trends in the hearing of disciplinary hearings and fitness to practice cases in the United Kingdom. But first, you may wish to complete the following self-study activities as these are designed to help you consolidate what you have learnt before moving on to consider the contents of Chap. 7.

Self-Study Activity

1. Write a 1,000-word essay which critically evaluates the strengths and weaknesses of the current plans pertaining to the implementation of revalidation in relation to their ability to identify and rehabilitate underperforming doctors.
2. Produce a 15-min PowerPoint presentation which critically considers if the current planned implementation of revalidation, particularly the role of the royal colleges in quality assuring the process, provides empirical support for the restratification thesis.

3. Write a 1,000-word essay which critically evaluates the development of periodic retesting arrangements for doctors from an international perspective. To what extent do recent developments reinforce that the medical profession is indeed undergoing a process of restratification as medical elites seek to retain some semblance of self-regulatory professional privileges, in the face of growing state intervention into medical regulation, by subjecting rank-and-file practitioners to greater peer surveillance and performance monitoring?

References

Boulay, C. (2000). Revalidation for doctors in the United Kingdom: The end or the beginning? *BMJ, 320*, 1490.

Burrage, M., & Torstendahl, R. (Eds.). (1990). *Professions in theory and practice*. London: Sage Publications.

Cain, F. E., Benjamin, R. M., & Thompson, J. N. (2005). Obstacles to maintaining licensure in the United States. *British Medical Journal, 350*, 1443–1445.

Calman, K. (1993). *Hospital doctors: Training for the future*. London: Department of Health.

Calman, K. (1994). *Continuing medical education*. London: Department of Health.

Cato, G. (2008). Relicensing, recertification and regulation. *Advances in Psychiatric Treatment, 14*, 1–2.

Catto, G. (2006). *GMC news 3*. London: GMC.

Chamberlain, J. M. (2009). Portfolio-based appraisal: Superficial or useful? *British Journal of Hospital Medicine, 70*(11), 176–177.

Charlton, R., Coomber, J., & Thistlewaite, J. E. (2011). Re-licensing of general practitioners using the current UK revalidation proposals: A cross-sectional study. *Postgraduate Medical Journal, 87*, 807–813.

Davies, M. (2007). *Medical self-regulation: Crisis and change*. Aldershot: Asj Publishing Limited.

Detsky, A., & Haylor, C. D. (2003). Canada's health system- reform delayed. *The New England Journal of Medicine, 349*, 804–810.

Duffey, F. D., & Zipes, D. P. (2004). The future of certification and recertification. *American Journal of Medicine, 117*, 140–144.

Freidson, E. (2001). *Professionalism: The third logic*. Cambridge: Polity Press.

General Medical Council. (1993). *Tomorrows doctors*. London: GMC.

General Medical Council. (2010a). *Revalidation update: Making revalidation work for doctors*. London: GMC.

General Medical Council. (2010b). *Revalidation: The way ahead*. London: GMC.

Gerace, R. V. (2003). *The shipman inquiry: International seminar statement by Dr Gerace*. London: The Shipman Inquiry Report.

House of Commons Health Committee. (2011a). *Revalidation of doctors: Fourth report*. London: The Stationary Office.

House of Commons Health Committee. (2011b). *Revalidation of doctors: General medical council's response to the committee's fourth report*. London: The Stationary Office.

Irvine, D. (2003). *The doctors tale: Professionalism and the public trust*. London: Radcliffe Medical Press.

Kelly, B. D. (2010). Changing governance, governing change: Medical regulation in Ireland. *International Journal of Medical Science, 179*, 5–7.

Lomdarts, M. J. M. H. (2003). External peer review by medical specialists in legal perspective. *European Journal of Health Law, 33*, 191–213.

Lupton, D. (2011). *Medicine as culture* (3rd ed.). London: Sage Publications.

Lynch, M. (2010). Commentary on Shepherd and Cameron and continuation of the revalidation debate. *Journal of Evaluation in Clinical Practice, 16*, 651–654.

Mekur, S., Elias, M., Long, M., & Mkee, M. (2010). Physician revalidation in Europe. *Clinical Medicine, 8*, 371–375.

Ministry of Social Affairs and Health. (2004). *Health care in Finland: Brochure II*. Helsinki: Ministry of Social Affairs and Health.

NHS Revalidation Support Team. (2009). *Strengthening NHS appraisal to support revalidation in England*. London: Department of Health.

Peck, C., McCall, M., McLaren, B., & Rotem, L. (2000). Continuing medical education and continuing professional development: International comparisons. *British Medical Journal, 320*, 432–435.

Pugh, M. (2003). Refining the paradigm: The transition from recertification to maintenance of certification. *Annals of Family Medicine, 1*, 56–58.

Rose, N. (1999). *Powers of freedom: Reframing political thought*. Cambridge: Cambridge University Press.

Smith, R. (1992). The GMC on performance: Professional self-regulation is on the line. *British Medical Journal, 304*, 1257–1258.

Smith, R. (1993). The end of the GMC? The government not the GMC is looking at underperforming doctors. *British Medical Journal, 307*, 957–958.

Smith, J. (2005). *Shipman: Final report*. London: Department of Health.

Stacey, M. (1992). *Regulating British medicine*. London: Wiley.

Stacey, M. (2000). The general medical council and professional self-regulation. In D. Gladstone (Ed.), *Regulating doctors* (pp. 45–65). London: Institute for the Study of Civil Society.

Starke, I., & Brownbridge, H. (2010). Revalidation: A royal college of physicians perspective. *Clinical Medicine, 10*, 321–322.

Swinkels, L. A. (1999). Reregistration of medical specialists in the Netherlands. *British Medical Journal, 319*, 1191–1192.

The Secretary of State for Health. (2007). *Trust, assurance and safety: The regulation of health professionals in the 21st century*. London: The Stationary Office.

Tito, F. (1996). *Compensation and professional indemnity in health care: Final report*. Sydney; Australian Government.

Chapter 7
Restratification and the Hearing of Fitness to Practice Cases

Abstract This chapter continues the exploration of recent developments in the regulation of the medical profession through the lens of the restratification thesis. As has been discussed in previous chapters, over the last two decades, medical regulation has had to change to become more open and accountable as health-care systems worldwide seek to better performance manage medical work. While at the same time, elite elements of the medical profession are themselves increasingly seeking to performance manage rank-and-file practitioner's activities as they respond to calls for increased regulatory transparency and accountability. In short, the restratification thesis argues that the medical profession is increasingly dividing into elite and rank-and-file segments as challenges to medical autonomy and self-regulatory privileges play out, and furthermore, medical elites are increasingly exploiting the specialist 'buffer zone' provided by their esoteric expertise as they seek to maintain some semblance of medical autonomy through subjecting rank-and-file doctors to greater peer surveillance and control mechanisms. Revalidation is arguably the latest example of this approach. But so are reforms to the hearing of fitness to practice cases. This chapter outlines contemporary developments in the complaint process as well as examines the latest statistical data pertaining to fitness to practice hearings. It highlights how the number of complaints has quadrupled in the last 15 years as well as how it appears that the GMC is adopting a more rigorous and punitive stance towards doctors accused of poor performance and/or unethical behaviour. The chapter also notes that male practitioners are more likely to receive a complaint than female practitioners (although this seems to be changing as more women join the profession). Additionally, older doctors and doctors who qualified outside of the UK are also more likely to come before the GMC fitness to practice panels. In outlining such matters, the chapter discusses how far recent developments in the handling of complaints provide empirical support for the restratification thesis. End-of-chapter self-study tasks are provided so the reader can engage in further study in relation to chapter contents.

J.M. Chamberlain, *The Sociology of Medical Regulation: An Introduction*,
DOI 10.1007/978-94-007-4896-5_7,

Introduction

This chapter is concerned with a key regulatory function which in the UK is overseen by the General Medical Council: the hearing of fitness to practice cases. An essential element of the principle of professional self-regulation is the possession of control over the process by which members of the medical profession are judged to have failed to live up to agreed clinical performance and/or ethical standards deemed appropriate for good medical practice. As Chap. 3 discussed, under the conditions of modern democratic governing imperatives, alongside the ever-shifting proclivities of the risk society, self-regulatory models of professional governance are prone to reoccurring crises of legitimacy because it is difficult to reliably control the actions of all professional members, and furthermore, when medical error does occur the charge can all too easily be made that self-regulation serves the interests of professionals themselves rather than the interests of their clients. This is certainly the conclusion drawn in relation to the GMC and its operation in the face of a series of high-profile medical malpractice cases, of which the Shipman case is but one example (Case 2011). Indeed, from the 1970s onwards, a series of high-profile medical malpractice cases came under the media spotlight and in doing so reinforced the legitimacy of the viewpoint that the GMC was by and large operating in favour of doctors and putting the interests of the medical profession before those of the public it claims to serve (Allsop 2002).

But it is perhaps the medical malpractice cases that came to light during the 1990s which finally signalled the beginning of the end for the traditional model of medical regulation. In particular, the organ-stripping scandal at Alder Hey Children's Hospital, the Bristol babies' inquiry, alongside the incompetent gynaecologist Rodney Ledward whose GMC fitness to practice case reinforced the medical bias which lay at the heart of the GMC's fitness to practice procedures, all can be said to have irreparably damaged the public image of the principle of medical self-regulation as it traditionally stood. Furthermore, as was outlined in Chap. 2, under the egis of the growth of the fluxing social conditions associated with the risk society, growing public and political doubt concerning the ability of professionals (and indeed any individual or group labelled as 'expert') to identify and control risk, means that, on one hand, we see an intensification in the use of evidence-based standardised protocols and guidelines to monitor and risk manage their activities, while on the other hand, we also see a shift in emphasis towards enabling individual members of society to negotiate and manage risk themselves (Beck 1992; Giddens, 1990, 1991, 1999). As both the risk society and governmentality perspectives remind us, an increasing focus on the promotion of individualism and personal responsibility lies at the centre of contemporary western societies. Indeed, we can see that over the last three decades doctors have become subject to a seemingly ever-increasing number of formal calculative regimes which seek to performance manage their work practices in order to better economise and risk manage occupational tasks while at the same time seeking to redefine the nature of the profession-patient relationship (Checkland et al. 2007; McDonald et al. 2008).

It should certainly therefore come as no surprise that these changes have been conceptualised in some quarters as suggesting there has been a shift towards the adoption of a risk-based approach to medical governance (i.e. Lloyd-Bostock and Hutter 2008). It has also been noted that the medical profession has collectively sought to respond to this changing regulatory landscape by internally reforming its self-regulatory institutions to modernise the training and regulation of doctors (Davies 2004). Similar to the introduction of clinical governance in the NHS, medical regulation now arguably seeks to promote a risk-averse working culture of transparency and accountability through the proactive use of clear performance standards and best-evidenced protocols and guidelines to inform decision-making processes (Irvine 2003).

Power (1997) emphasises the enormous impact of the contemporary trend in all spheres of western societies towards Audit in all its guises – with its economic cost-benefit concern with transparent accountability and standardisation – particularly for judging the activities of experts. In part then medicine is caught up within broader epoch-making social changes which to some degree are not of its own making. After all, as the governmentality perspective reminds us, there has been a societal wide shift in how good governance is conceptualised and enacted within neo-liberal nation-states. Nevertheless, it must be acknowledged that medical elites have been notoriously slow to reform the medical club while the fact remains that the historical evidence betrays the fact that there has long been a certain medical bias within the operation of the GMC (Stacey 2000), particularly in terms of its handling of fitness to practice cases. Consequently, the emergence of calls to reform medical regulation to make it more open and accountable have been particularly concerned over the last decade, especially after the Shipman case, with rectifying this state of affairs and ensuring the patient interest lies at the heart of GMC activity. Yet, as Chaps. 4 and 5 both outlined, the key problem faced by any intervention into the field of medical regulation is that it needs to take into account the fact that at some level the cooperation of the possessors of medical expertise is essential to the continued functioning of medical regulation (Chamberlain 2009).

There can be no return to the closed shop era of club government. It is undoubtedly the case that the 2088 Health and Social Care Act led to the GMC being reformed by the state to become, at least on the surface, an independent regulatory agency which must now operate under the banner of 'stakeholder regulation', called as such because both the public and other professionals must be involved as key partners in the regulatory process, while at the same time becoming subject to regulatory oversight from state-backed agencies such as the Council for Healthcare Regulatory Excellence (Chamberlain 2010a). However, as Waring et al. (2010:551) note, *'many regulatory activities continue to rely upon members of the profession in exercising their professional skill and judgment'*. Revalidation is, of course, one of these activities. The hearing of fitness to practice cases is undoubtedly another. Smith (2005), in her report on the Shipman case, argued persuasively for the need for an independent adjudicator in fitness to practice cases. In response, the 2008 Health and Social Care Act established the Office of Health Professions Adjudicator (OHPA) to take over the role of the GMC in the adjudication of fitness to practice cases. The stated aim of this change was to enhance impartiality and the independence of the fitness to practice hearing process within the health-care professions (Department of Health 2009). OHPA became a legal entity in January 2010. But perhaps unsurprisingly in the current economic climate given the GMC like many other professional regulators is largely a self-funding body, in the summer of 2010, the UK government concluded that it is not persuaded of the need to introduce another regulatory body to take over the role of adjudicator in fitness to practice cases (Department of Health 2010).

The governmental focus for now has moved to ensuring that GMC reform continues to enforce a shift towards a rigorous and fair complaint and fitness to practice adjudication process. Possible options voiced for consultation include a greater focus on the use of rehabilitative measures within the complaints system when concerns about a doctor's clinical performance exist, alongside the development of a more streamlined in-house tribunal system, headed by an independent president, who would be responsible for overseeing fitness to practice cases separately from the complaint receipt and management process, the handling of which would be retained at a day-to-day level by the GMC (Department of Health 2010). Only time will tell what the next steps in the reform of the regulation of doctors in the UK will be. But one thing is certain; although some progress seems to have been made in opening up the medical club, the contemporary regulatory situation is highly complex and it is not simply a matter of medical autonomy being in decline. Not least of all because the proactive participation of medical elites in regulatory reform is essential to achieving a modernised regulatory regime. Bearing this in mind in this chapter, we will explore contemporary reforms in the hearing of fitness to practice cases. In doing so, this chapter aims to highlight current trends in the management of complaints. It does this to ascertain if rank-and-file doctors are indeed coming under greater peer surveillance and control in an open and transparent manner and in the process becoming increasingly subject to formal rehabilitative or punitive action when their fitness to practise is called into question.

This chapter will also consider what the key consequences of this state of affairs may be for both patients and rank-and-file members of the profession. One of the key aims of this chapter is then to ascertain the validity of the restratification thesis in relation to the handling of fitness to practice cases. The removal of OHPA does lend some immediate legitimacy to the restratification position. But to more fully examine the situation, we must look at how complaints are handled by the GMC. To begin to do this, it is necessary to examine the role of the GMC in relation to the medical register.

The Medical Register and the Changing Medical Profession

When the GMC was first founded in 1858, the key regulatory goals for the medical profession at large were twofold: first, to establish a state recognised and hence legally underwritten register of approved medical practitioners and, second, for control over entry onto and exit from this register to primarily lie in the hands of the medical profession via representation on the GMC board by elite medical institutions, including most notably members drawn from university medical schools and the royal colleges. Put simply, after the 1858 Medical Act, nobody could get onto the medical register unless they had satisfied the minimum entry performance and ethical requirements, as set by the medical schools and royal colleges, and similarly, nobody could be removed from the medical register without a protracted fitness to practice process being completed. In 2011, there were 239,270 doctors on the GMC register (in 2001, there were 205,829 registered doctors, hence this data reinforces that the number of doctors in the UK has risen considerably over the last decade). This means that the UK has one doctor for every 365 people in the population. This compares to the USA, which has one doctor for every 375 of its population (World Health Organization 2011). 139, 381 (58%) of these 2011 registered doctors are male and 99,889 (42%) female.

The most significant demographic change in the medical profession over the past 10 years has been the growing number of female doctors. Men may still outnumber women but, based on current trends, it has been predicted that the women will become the majority of doctors in the National Health Service (NHS) in England by 2022 (Elston 2009). The changing gender balance is particularly pronounced amongst UK qualified doctors. Indeed, in the UK, the number of female doctors has risen by nearly 37% since 2001 and at the same time the number of male doctors has fallen by almost 8% (GMC 2011). The proportion of female members of the medical profession has similarly risen over the last two decades across many other developed countries (Deech 2009). While just as importantly it should also be noted that in 2011 48% of registered doctors described themselves as white and 26% of registered doctors described themselves as Black and Minority Ethnic (BME). However, ethnicity data for 26% of doctors was missing (GMC 2011). Nevertheless, in spite of this statistical gap, when taken alongside data pertaining to gender such figures reinforce that medicine as a profession is rapidly changing from its traditional predominately white male make-up.

All doctors must pay an annual fee to stay on the medical register. In 2011, the fee was £420, with a 50% discount for doctors earning under £26,000. Payment of registration fees is made even more essential by the fact that only registered medical practitioners may occupy medical posts within the National Health Service. The most common types of GMC registration are provisional and full. All graduating UK medical students obtain provisional registration as they enter employment in what was previously called the preregistration house officer year and is now referred to as the foundation year (or F1 as it is also known). This is converted into full registration on satisfactory completion of F1 and the now junior doctors enter specialist training (also referred to as foundation year 2 or F2). F2 and post-F2 higher specialist training is overseen by the GMC operating in tandem with one of 61 medical or surgical specialties and, within these, 34 approved sub-specialties. In the past, a third type of registration, limited registration, was granted to foreign graduates who had completed the Professional and Linguistic Assessment Board examination but required a period of work in the UK before they could obtain full registration.

However, limited registration was abolished in 2007, and now international medical graduates can apply for provisional or full registration depending on their level of work experience. All international (i.e. non-UK trained) medical graduates have to meet the GMC's requirement for knowledge and skills and for English language. The GMC administers what is called the Professional and Linguistic Assessment Board test (PLAB) which has to be sat by non-European Union overseas doctors before they may practice medicine in the UK. Oversees doctors are required, by law, to demonstrate that they possess the necessary knowledge of English. In 2010, the GMC reviewed the standard required from the language test and, after far reaching consultation with a range of stakeholders, increased the score required (GMC 2011).

For UK medical graduates, although a national exist examination, overseen by the GMC, has been discussed, entry to the medical register is de facto controlled by medical elites, in the form of the medical schools and the royal colleges, who respectively oversee the examinations and performance tests necessary for initial undergraduate degree graduation and subsequent postgraduate specialist progression. In contrast to this, although once solely the preserve of medical members, exit from the medical register over the last decade has come under greater non-medical scrutiny in the form of lay membership on fitness to practice panels as well as through regulatory oversight from the Council for Healthcare Regulatory Excellence. As of 2011, there were 146 medical and 119 lay fitness to practice panel members, dealing with a range of cases that are primarily concerned with practitioner's clinical competence and/or ethical probity. Such figures reinforce that the medical club has opened up to accept non-medical members. However, a cursory glance at the members list available online via the GMC website (http://www.gmc-uk.org/) shows that lay members tend to come from middle-class professional backgrounds (i.e. retired teachers, solicitors and even other health- and social-care professionals such as nurses and social workers). It has been pointed out before that the development of modern regulatory regimes requires the GMC (and indeed all professional regulators) move beyond the current tendency for lay members to be drawn from the 'great and the good' within society, as this potentially neglects the diverse range of social groups present within modern, pluralistic, multicultural western nation-states (Allsop 2002).

In conclusion, taken together, the current developments occurring in terms of the shifting make-up of the medical profession are clearly beginning to be reflected in the make-up of the GMC. Steps have been taken over the last decade to ensure lay members are included in all aspects of the regulatory process. As well as to ensure that female and ethnic minority groups have their voices heard as well. Yet entry onto and exit from the medical register still heavily involves medical elites in the form of medical schools and the royal colleges. Clearly, lay and other health-care professions are playing an increasing role in medical regulation. But it needs to be remembered that this involvement is limited insofar that peer appraisal of the clinical performance of a medical student, junior doctor, mid-career grade doctor, hospital consultant or general practitioner is essential if regulatory quality assurance mechanisms are to work. What we are seeing, therefore, is perhaps best described as the dilution and rolling back of medical power from within the GMC and its transfer into the medical schools and royal colleges. For at the same time as instigating regulatory change there is an increasing recognition on behalf of the state of the limitations of non-medical judgement when it comes to dealing with performance related matters. With the result that medical elites, in the form of the royal colleges and medical schools, are playing an increasing role in delivering the new regulatory agenda, in doing so, they are arguably acting to protect self-regulatory privileges, albeit under the egis of a new, more open, inclusive and transparent, regulatory regime.

Such developments, at least in principle, reinforce the continued legitimacy neo-Weberian model and the associated restratification thesis. Not least of all because, as Chaps. 2, 3 and 4 all discussed in some detail, occupational control over educational credentials and the primacy of peer judgements regarding clinical performance and competence together lie at the centre of the question of how best to regulate esoteric forms of expertise (Freidson 2001). In short, medical autonomy may have been curtailed, it may even be subject to more formal mechanisms of surveillance and control which incorporate both medical and non-medical expertise within them, but in the final analysis doctors still possess a significant degree of individual and collective control over matters of training, practice and discipline, particularly when compared to other occupational groups. Nevertheless, the need for standards-driven transparency and accountability does mean that just as medical elites move to place rank-and-file members under greater peer appraisal and performance management, so they themselves are becoming subject to audit and performance appraisal from state-backed agencies. In the rest of this chapter, we will explore the impact of this for the management of complaints. But first it is necessary to outline how the GMC complaint and fitness to practice panel process operates.

Handling Complaints

As already noted, most doctors working in the UK specialise in an area of practice once they have completed their initial education and training. The GMC currently approves 61 specialties (and 34 sub-specialities). The largest speciality is general practice, followed by (in order) anaesthetics, psychiatry, paediatrics, obstetrics and gynaecology, general surgery, trauma and orthopaedic surgery, clinical radiology, emergency medicine, geriatric medicine and cardiology. The GMC works closely with the royal colleges to sets the standards for training and the curricula and assessment systems (including examinations) for these specialties. However, no matter how rigorous a doctor's training, or how well a regulatory body acts to ensure quality assurance standards are maintained, the fact of the matter is that sometimes medical errors and mistakes will, and do, occur. Additionally, not only are doctors subject to the same physical and mental stresses and strains as the general population, unfortunately they sometimes do act in an inappropriate manner towards patients, that is, making sexual advances. In short, they are occasions when a doctor's clinical competence or ethnic probity is called into question. In such instances, the GMC's regulatory role is laid out by the 1983 Medical Act, as amended by the 2008 Health and Social Care Act. The GMC is responsible for removing doctors from this register by dealing firmly and fairly with medical practitioners whose fitness to practise has been questioned whether this be a matter of clinical competence, ill health or ethical probity (Chamberlain 2010a). The total number of written complaints about hospital and community health services in the 2010–2011 year was 89,139 (Department of Health 2011). In terms of professional and occupational groupings, the total written complaints figure of 89,139 breaks down as follows: medical (including surgical) 39,981; dental (including surgical) 908; professions supplementary to medicine 4,056; nursing, midwifery and health visiting 19,111; scientific, technical and professional 1,167; ambulance crews (including paramedics) 2,541; maintenance and ancillary staff 1,014; administrative staff 10,246; and other 10,115. The NHS possesses various mechanisms for dealing with complaints which operate at both local and national levels.

It is important to note that the GMC does not deal with complaints against NHS systems. This said, the GMC may well deal with complaints against individual medical practitioners which illustrate broader system failings within the NHS. Such as, for example, when a working culture of mutual secrecy between health-care practitioners (and even management) within a hospital unit has stopped concerns about medical staff being voiced. The inquiries into events at Bristol Royal Infirmary in the 1990s, at Maidstone and Tunbridge Wells in 2005–2006 and more recently at the Mid Staffordshire NHS Foundation Trust, have all exposed individual clinical failings of one kind or another, but each one has also demonstrated fundamental flaws in the way care was organised and the culture of institutions concerned (Department of Health 2000; The Bristol Royal Infirmary Inquiry 2001, Healthcare Commission 2007; Francis 2010).

Importantly, the GMC also does not arrange for complainants to receive an apology, an explanation of what happened, or provide help and support for compensation claims. Such matters form part of the broader NHS complaints system. Indeed, it is important to be clear that the GMC is one of a number of bodies which deal with complaints against medical practitioners (and indeed health-care and administrative NHS staff). NHS Hospital Trusts, Primary Care Trusts, alongside the National Clinical Assessment Service, the Healthcare Commission and the Parliamentary and Health Service Ombudsman, are all important points of contact for dealing with medical malpractice and patient complaints. However, and most importantly, the GMC remains the only body able to remove a doctor from the medical register and therefore stop them from practising medicine in the UK (Stacey 2000). The GMC only deals with complaints that call into question a doctor's fitness to practise (GMC 2004a). The GMC's operational bounds, processes and imperatives are clearly stated via statutory legislation passed by the UK parliament and only parliament can change its operation. Under Section 35 C(2) of the Medical Act (1983), alongside the guidance to good practice provided in its document Good Medical Practice (2009), the GMC focuses upon complaints which highlight instances where a doctor has made serious or repeated mistakes in carrying out medical procedures or in diagnosis (i.e. by prescribing drugs in a dangerous way), has not examined a patient properly or responded appropriately to their medical need, has committed fraud, dishonesty or serious breaches of a patient confidentiality and, finally, has received a criminal conviction or has developed a physical and/or mental health issue (Chamberlain 2010a).

The GMC guidance in Good Medical Practice (2009) specifically requires doctors make the care of their patient their first concern via the use of the following thematic headings which together are taken as defining the duties of a doctor: to protect and promote the health of patients and the public, to provide a good standard of practice and care, to keep professional knowledge and skills up to date, to recognise and work within the limits of competence, to work with colleagues in the ways that best serve patients' interests, to treat patients as individuals and respect their dignity; to treat patients politely and considerately, to respect patients' right to confidentiality, to work in partnership with patients, to listen to patients and respond to their concerns and preferences, to give patients the information they want or need in a way they can understand, to respect patients' right to reach decisions with a doctor about their treatment and care, to support patients in caring for themselves to improve and maintain their health, to be honest and open and act with integrity, to act without delay if a doctor has good reason to believe that they or a colleague may be putting patients at risk, to never discriminate unfairly against patients or colleagues and to never abuse patients' trust in their doctor or the public's trust in the profession. Guidance is provided on each of these thematic categories which provides standards for doctors to follow and taken together they form part of the GMC's commitment to promoting good medical professionalism as well as transparent and accountable regulatory quality assurance frameworks (GMC 2009).

From Triage to Investigation and Adjudication

As a result of the Shipman inquiry in 2004, the GMC internally reformed its fitness to practice procedures. Traditionally, cases were dealt with by three separate committees: health, conduct and performance. The health committee was concerned with matters relating to a doctor's physical and mental health. The conduct committee was concerned with a doctor's ethical probity. The performance committee was concerned with a doctor's clinical performance. Although broadly operationally separate from one another, in reality, the activity of each committee overlapped. This often led to prolonged delays in the hearing of cases, and in an effort to speed up and simplify the process in 2004 the committees were combined and the complaint management process divided into two separate stages: investigation and adjudication. During the investigative stage what is known as the initial triage process involves making an initial decision as to whether or not to proceed with the case (GMC 2004a). Some complaints received are clearly outside of the GMC's remit. For example, a complaint may not be concerned with an individual medical practitioner. If necessary, the GMC will refer the matter to the doctors' employer so local procedures can be used if needed to respond to it. If the initial information points towards a criminal conviction, then the matter will be immediately referred to a fitness to practice panel for adjudication (GMC 2004a). If the triage process confirms that the complaint requires further consideration, the GMC will proceed to the full investigative stage. Here, the GMC will disclose the complaint to the doctor in question and their employer to ensure a complete picture of the doctor's practice can be obtained.

Cases can be placed in one of two streams. Stream two complaints contain cases whereby the information received is not considered to be serious, but would be of concern if it were part of a wider pattern of behaviour or practice, so the GMC will contact the doctor's employers to ask for further information about the doctor's practice. A decision will then be made as to whether there needs to be an investigation under stream one conditions. Stream one complaints raise serious concerns about a doctor's fitness to practise and/or health/ethical probity. All cases are overseen by two case examiners, one of whom is a lay GMC member and one a medical GMC member. Witness statements and supportive material will be collected and analysed, including copies of patient medical records or other formal documentary material (i.e. employer reports, medical audit outcomes, surgical operation outcomes and so on). Where there is a concern with performance or health, appropriate tests will be completed at this stage (Etheridge et al. 2009). The GMC has at its disposal a range of health specialists, including psychiatrists and alcohol and substance abuse specialists. The investigation period concludes with either no further action been taken, a warning being issued, a practitioner agreeing to what are referred to as undertakings or a case being referred to a fitness to practice panel for adjudication (GMC 2010). Undertakings are an enforceable agreement between the GMC and a doctor. Their duration can last for a maximum period of 3 years. They might include restrictions on a doctor's future practice or behaviour, as well as the requirement that they commit to having medical supervision or retraining. Retraining, clinical competence and supervisory requirements will be overseen by a number of parties to ensure effective risk management. All undertakings are regularly reviewed by the GMC, operating in liaison with a doctor's employer as well as postgraduate and specialist medical training providers (i.e. the royal colleges). In comparison, a warning occurs when there is a significant concern about a doctors' practice, but imposing restrictions on their practice is not held to be necessary.

The adjudication stage involves a formal hearing of a case by a fitness to practice panel. The panel is made up of medical and non-medical lay members. If needed, the panel will be advised by a specialist health or performance advisor. They are five main outcomes of a fitness to practice panel meeting: no further action, giving a doctor a formal warning, putting restrictions of a doctor's professional practice (i.e. imposing supervision or requiring they undertake further training), suspending a doctor from the medical register so they may not practise for a given period of time and, finally, erasing a doctor from the medical register. It is the intention of the GMC when they erase a doctor from the medical register that this normally will be for life. A doctor has 28 days to appeal against a decision which they lodge at the High Court. Since 2005 all GMC fitness to practice decisions have been reviewed by the Council for Healthcare Regulatory Excellence. Under section 29 of the National

Health Service Reform and Health Care Professions Act (2002) the Council can refer a decision to a High Court for review if it considers a GMC decision to be unduly lenient. The Council for Healthcare Regulatory Excellence forwarded 4 such cases in 2005, 6 in 2006, 0 in 2007, 1 in 2008, 1 in 2009 and 0 in 2010 (Council for Healthcare Regulatory Excellence 2005, 2006, 2007, 2008, 2009, 2010a).

This reduction in referrals could be held to reflect an increasingly rigorous stance on behalf of the GMC towards fitness to practice cases (Allsop 2006). There is certainly a growing perception within the medical profession that the GMC is far less tolerant of infractions than it was previously (Dyer 2010a, b, c). Many practitioners are concerned with what they perceive to be the increasing politicisation of the operation of the GMC (Chamberlain 2009, 2010a). This is particularly the case since the 2008 Health and Social Care Act changed the level of evidence required to remove a doctor from the medical register. For example, during each stage of the complaint process a case must pass what is called the realistic prospect test (Case 2011). Simply put, this means that the allegations will only proceed if there is a realistic prospect of establishing that a medical practitioner's fitness to practise can be called into question to such a degree that justifies the GMC taking action on their registration status (GMC 2004a). Under the Health and Social Care Act (2008), the level of evidence required to secure a guilty verdict has been reduced from a criminal (absolute) to civil (on the balance of probabilities) standard. This change was ultimately a political decision made by the Labour government through passing an act of parliament after consultation with key stakeholders, that is, doctors, medical malpractice insurers and lawyers, patient rights groups and NHS representatives. The decision was made in the face of heavy lobbying against the measure from medical elites and the medical profession at large via the British Medical Association (BMA) as the main professional union for doctors. Nevertheless, the change was justified by the state on the grounds that the GMC has often in the past been unable to remove a doctor from the medical register, even when doubt existed over their clinical performance, because the level of evidence required to do so was too high (see Irvine 2003; Allsop 2006).

After the Shipman case, many members of the profession agreed changes were needed to update the organisation and working culture of the GMC so underperforming doctors could be more easily stopped from continuing to practise. But it has also been argued by the British Medical Association that moves to reduce the level of evidence needed to remove a doctor from the medical register had become bound up with of a wider politically motivated and unrealistic tendency, on behalf of the regulatory state and NHS management, to seek to minimise clinical risk and cost by turning medical work into a series of routine step-by-step rules and procedures against which individual clinician performance can be measured and judged (Chamberlain 2009). They are clearly several issues here relating to contemporary developments in medical regulation which require further discussion. Including, perhaps most importantly, the matter of if the opening up of the GMC and the adoption of more rationalistic measures have brought about a significant decline in medical autonomy. To answer such questions, it is necessary to explore the statistics relating to GMC activity and as part of this identify if there had been a substantial rise in the number of doctors removed from the medical register as a result of changes in the level of evidence needed to pass the realistic prospect test. Hence, the next section of this chapter will detail this information before moving on to discuss its consequences.

Trends in Complaints by Source and Practitioner Characteristics

Having outlined the process by which a fitness to practice complaint proceeds, we will now focus on data pertaining to the hearing of fitness to practice cases by the GMC. This chapter will draw on the activity data published by the GMC for between 2006 and 2010 with data from earlier years being discussed where possible (see GMC 2000, 2001, 2002, 2003b, 2004c, 2010, 2011). To begin with, it is important to recognise that although the figures discussed do descriptively illustrate the operation of the GMC they should not be taken as a representation of its total activity for each calendar year. Not least of all because, as would be expected, the nature of the complaint process is such that a complaint received in 2009 may not reach resolution until 2010. The GMC does seek to resolve complaints as quickly as it can. However, a range of factors, including the need to collect and analyse witness statements, operational data and specialist reports while also adhering to strictly defined legal processes, can and do often prolong the case management and resolution process. This proviso to one side, having year-on-year comparative data does allow for descriptive statistical trends to emerge, as this chapter will now turn to discuss. First, let us begin with the total number of complaints the GMC receives. The total number of complaints received by the GMC for between 1999 and 2009 are as follows: 1995, $n = 1{,}503$; 1998, $n = 3{,}066$; 1999, $n = 3{,}001$; 2000, $n = 4{,}470$; 2001, $n = 4{,}504$; 2002. $n = 3{,}937$; 2003, $n = 3{,}962$; 2004, $n = 4{,}005$; 2005, $n = 4{,}128$; 2006, $n = 2{,}788$; 2007, $n = 4{,}118$; 2008, $n = 4{,}166$; 2009, $n = 4{,}722$; and 2010, $n = 7{,}153$ (Source: GMC 2000, 2001, 2002, 2003a, b, 2004b, c, 2010, 2011). These figures show that, aside from 2006 when the number of complaints for some reason reduced sharply, the number of complaints received by the GMC has roughly quadrupled over the last 15 years. That is from a baseline of 1,503 in 1995 to 7,153 in 2010. Yet it should be remembered here that 7,153 complaints represent 3% of all medical practitioners currently on the GMC register ($n = 239{,}270$). Clearly then we are talking about a small number of doctors and such statistics reinforce the commonly held position that in general public satisfaction with the medical profession remains buoyant. The dip in complaints in 2006 cannot be attributed to any major change in the organisation or role of the GMC during this year, so it may well simply be a statistical aberration, as does happen sometimes when dealing with longitudinal data. It appears that complaints doubled between the mid and late 1990s, with the number trebling into the beginning of the new millennium, before evening off slightly (aside from in 2006) until increasing back up again in 2009 and again in 2010.

It is too early to tell yet if the jumps in the number of complaints in the last 2 years will persist or level off. Yet it is clear that the available figures do seem to reinforce the validity of the view that in the last two decades there has been an increase in the questioning of medical authority and autonomy, with the result that individuals are more likely to complain about their doctor and/or the treatment they have received (Nettleton 2006). Bound up with this may well be the fact that high-profile medical malpractice cases, such as the respective Bristol Infirmary and Shipman cases, have significantly raised the profile of the GMC in the eyes of the news media and general public, with the result that the number of complaints it receives has increased (Chamberlain 2010b). Amongst their other duties medical directors oversee medical audit and quality assurance mechanisms and act as points of liaison between the GMC, NHS management, rank-and-file practitioners and patients when complaints arise. Recent research conducted with a large number of medical directors suggests that not only does the rise in fitness to practice complaints reflect an increasingly litigious culture within society, as well as more generally changing patient attitudes towards doctors and a concurrent willingness to complain, but also doctors themselves seem to be more willing to refer colleagues if they have a concern for patient safety (Growth from Knowledge 2011). These points bring us to the issue of the statistical data pertaining to the source of complaints.

The analysis of the source of complaint was complicated by the fact that the GMC reporting of complaints received by each main category – that is, members of the public, a fellow doctor, a person acting in a public capacity such as a nurse or social worker – is different in the data source documents for between 2000 and 2005 (GMC 2000, 2001, 2004a, b, c) than for the 2006 to 2010 years (GMC 2010, 2011). But it was possible to combine the categories into a simple public/other dichotomy to help paint a broad picture of patterns in the source of complaints received by the GMC. The 'Other' category here includes complaints from medical practitioners, other health- and social-care professionals and NHS management/administrative/support staff. The source of complaint data for each year by public/other categories is as follows: 2000, Public 73% Other 27%; 2001, Public 75% Other 25%; 2002, Public 76% Other 24%; 2003, Public 77% Other 23%; 2004, Public 74% Other 26%; 2005, Public 73% Other 27%; 2006, Public 66% Other 34%; 2007, Public 70% Other 30%; 2008, Public 68% Other 32%; 2009, Public 64% Other 36%; and 2010 Public 63% Other 37% (Source: GMC 2000, 2001, 2002, 2003a, b, 2004b, c, 2010, 2011). This data reinforces that the majority of complaints received come from the general public. Yet there has also clearly been a gradual increase in complaints from other sources in recent times. This may well reflect the fact that the GMC has recently taken a more proactive stance towards working with local NHS employers and private health-care providers as it seeks to promote a working culture which encourages complainants to come forward with their concerns without fearing negative consequences for their career, particularly as this has been recognised as a key issue in the past (Department of Health 2009).

It is also important to note here that it is only since 2006 that the GMC has systematically collected and stored data on the specific types of concerns brought to it. This in itself reflects the closed shop mentality which was at play within the medical club until relatively recently. The three persistent types of complaint relate to poor clinical performance and/or care, disregard to patients and poor communication with patients (GMC 2011). Under the GMC remit complaints about poor working relationships with patients are more likely to be referred back to NHS Trusts to be handled locally, quite possibly with rehabilitative action being taken in relation to doctor-patient communication training or completion of patient rights, diversity and/or equality training programmes. Indeed, ethical probity and quality of clinical care remain chief concerns for the GMC, with these accounting for 72.9% of all erasures from the medical register in 2010 (GMC 2011). Nevertheless, the GMC has in the last decade become increasingly concerned about poor communication with patients and alongside the royal colleges has taken steps to seek to promote communication skills within undergraduate and postgraduate training. It is interesting to note here that the GMC is more likely to receive complaints about older than younger doctors, particularly older male doctors. This may well be because younger doctors are still in training and so are being overseen by postgraduate deaneries and the royal colleges.

Similar conclusions about age and performance were drawn earlier this year by the National Clinical Assessment Service (NCAS). This body deals at a local level with doctor and other health-care practitioner retraining when concerns about performance are raised. Their data relating to 5,600 referrals between 2001 and 2010 revealed higher levels of concern relating to clinical competence for older doctors; the likelihood of referral to NCAS beyond age 60 is about seven times the likelihood below age 40 (National Clinical Assessment Service 2011). It has been argued that this pattern seems to provide evidence for the argument that practitioners in the later stages of their careers need additional educational support. Certainly, although the majority of older doctors work effectively, this data does underline the need for all medical practitioners to keep up to date with latest developments in their discipline, and it will be interesting to see if and how this data changes with the introduction of revalidation nationally from 2012 onwards.

Having broken down the complaint data by source, type and age of doctor, the next step in our analysis is to break down the initial complaints made against doctors against their ethnicity and gender. It was not possible to obtain data pertaining to complaints received in relation to ethnicity prior to 2006 from the documents available via the GMC website. It is important to note that the GMC did not routinely collect data pertaining to ethnicity prior to the introduction of a new electronic recording system in 2005. Furthermore, during 2007and 2008, it '*undertook a major exercise to improve the quality and coverage of its ethnicity data*' (Humphrey et al. 2009:19). In part, this has been because of growing concerns over possible discrimination and racism, as doctors who qualified overseas and subsequently came to practise in the UK seem to be at higher risk of being referred to the GMC as well as having high-impact fitness to practice decisions made against them, that is, having limitations placed upon their practice or being struck off the medical register (Allen 2000; West et al. 2006).

Humphrey et al. (2009) in their review of available data for between 2006 and 2008 found that UK doctors from ethnic minorities were not at greater risk of being subject to high-impact decisions or being struck off the medical register; however, overseas doctors did seem to be. It should be noted that Humphrey et al. (2009) were cautious concerning their findings due to the limited data available to them. Nevertheless, the GMC data available concerning the percentage breakdown of initial complaints by ethnicity shows relative consistency in the complaints by race and ethnicity category over the last 4 years: (a) ethnicity Asian or Asian British 2006, 18%; 2007, 17%; 2008, 18%; 2009, 19%; and 2010, 19%; (b) Black or Black British 2006, 3%; 2007, 3%; 2008, 3%; 2009, 4%; and 2010, 3%; (c) mixed 2006, 1%; 2007, 1%; 2008, 2%; 2009, 1%; and 2010, 2%; (d) other ethnic groups 2006, 3%; 2007, 3%; 2008, 3%; 2009, 3%; and 2010, 4%; (e) unspecified 2006, 29%; 2007, 29%; 2008, 26%; 2009, 27%; 2010, 24%; and (f) White 2006, 46%; 2007, 47%; 2008, 48%; 2009, 45%; and 2010, 48% (Source: GMC 2010, 2011). From this data, it seems to be the case that Asian or Asian British ethnic minorities are overrepresented in terms of complaints made against them. It is not known how many of those individuals classified as Asian or another ethnic minority by the GMC data set come from overseas. 2001 census data shows that 7.9% of the UK population belong to an ethnic minority, and that the Asian or Asian British category accounts for 4% of the population as well as 50% of all ethnic minorities (Office of National Statistics 2001). The available data reinforces the need for a doctor's ethnicity to be as far as possible recorded when a complaint is received as currently unspecified remains a major response category. This may be expected somewhat given that it is not always possible for a complainant to know a doctors ethnicity. But it would nevertheless be expected that the unspecified category would decline if the GMC were to adopt a more proactive (and possibly more retrospective) stance on the recording of ethnicity data in relation to initial complaints received. Although perhaps an unavoidable factor at play here may well be, as will be discussed in more detail shortly, that the majority of complaints do not make it past the initial triage stage as this obviously complicates the retrospective collection of ethnicity data in relation to initial complaints received. Nevertheless, the GMC clearly needs to take steps to investigate the issue of the overrepresentation of oversees doctors in fitness to practice proceedings to ensure its proceedings are fair and non-discriminatory.

Next it is necessary to break down complaints by a medical practitioner's gender. It was possible to extract from GMC documents data pertaining to complaints against male and female doctors for 2002 onwards. However, it was not possible to obtain data for the 2005 year. Similar to ethnicity, reliable data has only become available relatively recently due to the recent review and computerisation of GMC records. The available data suggests that complaints are more likely to be made about male doctors than female doctors, although it does appear to be the case that complaints against female doctors rose slightly over the last 4 years. The breakdown by gender per year from 2002 was as follows: 2002, male 83% female 17%; 2003, male 82% female 18%; 2004, male 81% female 19%; 2005, unknown; 2006, male 79% female 21%; 2007, male 81% female; 2008, male 81% female 19%; 2009, male 78% female 22%; and 2010, male 75% female 25% (Source: GMC 2000, 2001, 2002, 2003a, b, 2004b, c, 2010, 2011). The greater emphasis on male doctors may well reflect the fact that the medical workforce has traditionally been male dominated and that the GMC caseload mix includes breaches of fitness to practise which are perhaps more commonly associated with male rather than female risk-taking behaviour, that is, improper sexual relationships with patients, criminal activity and substance/alcohol abuse (Stacey 1992, 2000). The proportion of female doctors in the profession has risen significantly in the last decade, with projections suggesting that by the early 2020s the majority of doctors will be female, which may help explain the slight proportionate rise in complaints against female doctors (Winyard 2009).

Latest analysis of fitness to practice panel data shows that while male doctors are more likely to be referred to a fitness to practice panel, men and women were equally likely to be erased from the register and a higher proportion of women than men were suspended (GMC 2011). International research has shown a similar disparity between male and female practitioners in terms of complaints (Taragin et al. 1992; Bratland and Hunskår 2006). Consequently, as the gender make-up of the medical workforce changes over the next decade it will be interesting to see if and how the compliant caseload of the GMC changes.

Trends in Fitness to Practice Hearing Activity

Having looked at the number of complaints made, it is now necessary to look at the figures relating to the progression of cases from the investigation and adjudication stages. Not least of all because this will allow us to ascertain if and how the GMC handling of complaints has changed over time. We already know the case load has increased over the last 15 years, but this raises the question of this has led to an increase in doctors being subject to punitive or rehabilitative measures or being struck off the medical register. Knowing this is important if we are to identify how far the working culture of the GMC has changed under the new regulatory regime. Yet it is necessary to begin this analysis by highlighting that it is not possible to identify with certainty statistical data prior to 2006. To some extent, this is down to the fact that it is only in the last few years that moves towards a working culture of regulatory transparency and accountability have led to an intensification in effort, on behalf of the GMC, to systematically collect and publish complaint and fitness to practice data. But perhaps the most important factor here is that available complaint data for between 2000 and 2005 suffers from the complication that the process by which the GMC handles complaints changed during this period as part of reforms introduced in light of the Shipman case.

As was mentioned previously, until 2004 the GMC's fitness to practice procedures were governed by separate legislation involving different committees concerned with three aspects of a doctors fitness to practise: health, conduct and performance. Leaving aside the fact that no electronic records were formally recorded and stored by the GMC until the beginning of the millennium, the available figures for the operation of each committee make it difficult to identify with certainly data pertaining to case outcomes. Not least of all because there was some natural overlap with the handling of cases by each committee. It should also be noted that some reports (e.g. for the 2004 year, see GMC 2004a, b, c) only break down the data for part of the year (in the case of the 2004 report for between January and October 2004), making it impossible to compare data year on year. Nevertheless, the available data for between 2006 and 2010 does reveal some interesting trends concerning the GMC handling of complaints.

Let us begin at the beginning and examine the data concerning what happened after each complaint was received by the GMC. For year-on-year comparative analysis purposes only the numeric data has also been broken down into the relative percentages for each complaint case action category based on the total number of complaints received during a year. In accordance with the GMC reports, complaint data can be broken down as follows (Source: GMC 2010, 2011):

Complaints received: 2006, 2,788 (100%); 2007, 4,118 (100%); 2008, 4,166 (100%); 2009, 4,722 (100%); and 2010: 7,153 (100%)
Concluded at triage: 2006, 1,970 (71%); 2007, 2,953 (71%); 2008, 2,872 (69%); 2009, 3,226 (68%); and 2010, 5,087 (71%)
Concluded at investigation: 2006, 472 (17%); 2007, 769 (19%); 2008, 658 (16%); 2009, 870 (18%); and 2010, 497 (7%)
Concluded (both): 2006, 88%; 2007, 90%; 2008, 85%; 2009, 86%; and 2010, 78%
Warning issued: 2006, 86 (3%); 2007, 159 (4%); 2008, 168 (4%); 2009, 212 (5%); and 2010, 183 (3%)
Undertakings agreed: 2006, 44 (1%); 2007, 40 (1%); 2008, 109 (2%); 2009, 95 (2%); and 2010, 102 (2%)
Advice: 2010 report only 458 (7%)
Referred for adjudication: 2006, 216 (8%); 2007, 196 (5%); 2008, 359 (9%); 2009, 319 (7%); and 2010: 314 (5%)

In pure numerical terms, the data shows that more warnings and rehabilitative undertakings occurred in the last 4 years than the first two, as well as that more cases are being referred for adjudication than previously. While the comparative data reveals some important consistencies in the GMC handling of complaints received and subsequent actions undertaken, that is, in spite differences in the number of complaints received each year. Indeed, the data shows that the majority of complaints are closed with no further action either at the initial triage stage (between 68 and 71% of complaints received over the 5-year period) or after the initial investigation has been completed (between 85 and 90% of all complaints received over the 5-year period). When disciplinary action is taken at this stage the doctor in question either agreed to rehabilitative undertakings (between 1 and 2% over the 5-year period) or has been issued with a written warning (between 3 and 5% over the 5-year period). Importantly then, although more complaints were referred for adjudication via a fitness to practice panel in 2010 than 2006, proportionally speaking only a relatively small percentage of complaints made it past the investigative stage (minimum 10% in 2007 and maximum 22% in 2010 over the 5-year period between 2006 and 2010). Finally, as already noted, in 2010, the GMC received complaints concerning 3%

(n = 7,153) of all doctors on the medical register (n = 239, 270) that 0.13% of all doctors on the medical register were referred to a fitness to practice panel for adjudication during that year (n = 314).

Having identified how complaints are handled by the GMC at the initial investigative stage, it is now necessary to detail the outcomes of cases heard at the adjudication stage by a fitness to practice panel. For year-on-year comparative purposes, the data has been broken down into relative percentages for each action category based on the total number of cases heard per year. In accordance with the GMC reports, complaint data can be broken down as follows (Source: GMC 2010, 2011):

Cases heard: 2006, 221 (100%); 2007, 257 (100%); 2008, 204 (100%); 2009, 270 (100%); and 2010, 326 (100%)

Impairment – no action: 2006, 8 (4%); 2007, 13 (5%); 2008, 4 (2%); 2009, 4 (1%); and 2010, 4 (1%)

No impairment – no action: 2006, 47 (21%); 2007, 36 (14%); 2008, 28 (14%); 2009, 44 (16%); and 2010, 65 (20%)

Voluntary erasure: 2006, 3 (1%); 2007, 2 (1%); 2008, 0 (0 %); 2009, 3 (1%); and 2010, 7 (2%)

Undertakings: 2006, 4 (2%); 2007, 4 (2%); 2008, 3 (1%); 2009, 3 (1%); and 2010, 5 (2%)

Reprimand: 2006, 1 (1%); 2007, 1 (1%); 2008, 0 (0%); 2009, 1 (1%); and 2010: 0 (0%)

Warning : 2006, 4 (6%); 2007, 8 (3%); 2008, 22 (11%); 2009, 22 (8%); and 2010, 29 (9%)

Conditions: 2006, 38 (17%); 2007, 55 (21%); 2008, 30 (15%); 2009, 48 (18%); and 2010, 37 (11%)

Suspension: 2006, 69 (31%); 2007, 78 (30%); 2008, 75 (37%); 2009, 77 (29%); and 2010, 106 (33%)

Erasure: 2006, 37 (17%); 2007, 60 (23%); 2008, 42 (20%); 2009, 68 (25%); and 2010, 73 (22%)

This statistical data reveals slightly less year-on-year consistency in the types of action taken at the adjudication stage than at the investigative stage. But it also reinforces that at the adjudication stage the hearing of complaints is more likely to result in high-impact decisions, such as conditions being placed on a doctors practice (between 15 and 21% over the 5-year period), suspension from the medical register (between 29 and 37% over the 5-year period) or erasure from the medical register (between 17 and 25% over the 5-year period). Relatively few doctors receive undertakings or warnings at adjudication stage: it seems that the most common outcome of the adjudication stage is either a high-impact decision or the decision that there was no impairment in a doctors' practice. There also appears in the last 5 years to have been an increase in the relative proportion of doctor's being erased from the medical register as well as a decline in the decision that there was no impairment in a doctor's practice. Nevertheless, it does not seem that the shift to a civil standard of evidential proof has resulted in an immediate and significant increase in doctor's being erased from the medical register, as was feared it would by some quarters of the profession (Chamberlain 2009). Finally, taken together, the investigation and adjudication data show that 0.13% of doctors on the medical register (n = 239, 270) were referred to a fitness to practice panel for adjudication in 2010 (n = 314) and 0.03% (n = 73) were erased from the medical register.

Fitness to Practice Activity and the Transformation of Medical Autonomy

The descriptive statistical data presented in this chapter highlights some key thematic trends in the handling of complaints by the GMC and the types of action which subsequently occur. In summary, the number of complaints has tripled since the mid-1990s, although it should be noted that the number of complaints remains low, that is, in 2010 the GMC received 7,153 complaints which represent 3% of all medical practitioners currently on the GMC register (n = 239,270). The data revealed that the GMC receives more complaints from members of the public than from other sources. This is perhaps unsurprising as it appears that the majority of complaints the GMC receives relate to issues to do with clinical care and/or poor attitudes towards, or communication with, patients. However, the number of complaints from other sources appears to be rising and this may well be illustrative of broader cultural change within both the NHS and the GMC. At least in terms of ensuring potential whistle blowers feel they can come forward without fear of personal repercussions, particularly in terms of ensuring a positive working atmosphere, good collegiate relationships as well as that subsequent career development pathways are not negatively affected when they come forward, as has often been the case in the past (Department of Health 2000, 2009).

The complaint data also shows that more complaints about male doctors than female doctors. This in all likelihood is a reflection of the fact that the GMC complaint caseload mix includes breaches of fitness to practise more commonly associated with male than female risk-taking behaviour, that is, improper sexual relationships with patients, criminal activity and substance/alcohol abuse (Staccy 1992, 2000). It will be interesting to see if and how the type of complaint changes as female members of the profession begin to predominate over the next decade. Additionally, it is important to note that although the majority of complaints are made against white doctors, it seems that Asian or Asian British doctors are over represented in terms of initial complaints received. But this finding must be treated with extreme caution, as the breakdown of ethnicity data for initial complaints is incomplete, while published research shows that doctors from an ethnic minority are not more likely to be subject to a high-impact decision, but it seems doctors who qualified overseas are (Allen 2000). This said, clearly it is important that this issue be explored further at both an NHS and GMC complaint level. While it is clearly necessary for the GMC to address this matter of the over representation of oversees doctors in complaint data and fitness to practice panel activity through ensuring that it is not acting discriminately and adheres to equal opportunities legislation.

One of the key themes emerging from the data is that although the numbers remain relatively small in terms of total complaints made, there can be no doubt that the data discussed reveals that more doctors are being subject to rehabilitative and disciplinary action by the GMC than previously has been the case. Importantly, however, the data also illustrates that the shift towards a civil standard of proof does not seem to have had an immediate large-scale impact on GMC fitness to practice activity, that is, significantly more doctors have not been struck off the medical register. But it does seem that more cases have been referred to the adjudication panel in the last 3 years than previously. It is perhaps too early to tell if the shift in the level of evidence required to meet the realistic prospective test will result in more complaints passing from the investigative to adjudication stage and more doctors being either struck off the medical register or subject to some form of disciplinary or rehabilitative action. But it does seem to be the case that the common perception amongst rank-and-file practitioners that the GMC has adopted a more punitive stance towards practitioners is correct.

This is also leads to an important point in relation to the restratification thesis. For the evidence presented in this chapter does lend further credence to the position that doctors are becoming subject to greater performance and control mechanisms. The operational culture of the GMC is undoubtedly changing. In the past the GMC has been accused of being self-serving, biased in favour of doctors, failing to protect patients, being overly secretive, as well as acting through expediency rather than principle (Chamberlain 2009). Smith, in her review as part of her analysis of the Shipman case, was particularly critical of the GMC and how it handles complaints (Smith 2005). Yet the growing emphasis being placed on taking rehabilitative or punitive action against doctors could be interpreted as providing evidence that an organisational and cultural shift towards a more risk-averse regulatory model as the GMC acts to regain public trust in its decision-making processes (Allsop 2006). What this reinforces is that both medical elites and other state-backed administrative and risk-management agencies (i.e. the promoters of clinical guidelines and best practice frameworks such as the National Centre for Clinical Excellence) are subjecting rank-and-file doctors to greater performance surveillance and correction. This is in line with the restratification thesis. Yet it also needs to be acknowledged that the GMC is no longer under the same medical-only form of professional control as it was previously. Rather, medical influence and control is now increasingly coming at a distance from the royal colleges and medical schools who must work within the more open, accountable and multidisciplinary operational culture emerging within the GMC. One then must imagine here that medical elites have retreated into their traditional power bases while at the same time seeking to utilise their specialist expertise to exert influence and control at a distance over regulatory mechanisms through their strategic control over standards and performance judgements in relation to training and quality assurance mechanisms. What this means in practice is that it is no longer appropriate to argue that the restratification of the medical profession into elite and file roles acts to preserve medical autonomy. Rather, medical autonomy has not been maintained in its traditional form (as the restratification thesis sometimes seems to imply), but instead medical autonomy is to some extent being transformed and recast to reflect the prevailing governing conditions of the risk society.

This new contemporary form of medical autonomy is a distinctive form of risk-aware governmentality in that it operates at a distance to ensure transparency and accountability through the use of codified performance standards, that is, as far as it is possible to do so when one is dealing with an esoteric form of applied expertise. Indeed, it is the tacit dimension of medical expertise which enables restratification to occur and medical autonomy to be appropriately recast so that doctors are still able to exercise a considerable degree of legitimate personal and collective judgement while also being increasingly subject to performance appraisal mechanisms. This state of affairs will be discussed in Chap. 8. But first let us turn to briefly consider what it may mean for both the profession and the public.

Medical Risk, Defensive Medicine and the Patient Voice

An important consequence of this shift towards risk-averse forms of medical governance perhaps comes most clearly to the foreground when the GMC's administratively robust approach towards the handling of complaints is considered. Under the conditions of the risk society professional regulation relies heavily on seemingly objective decision-making processes where codified forms of knowledge are used to prescribe best-evidenced judgemental norms surrounding what constitutes appropriate action in a given situation (Lloyd-Bostock and Hutter 2008). Hence, we have seen the emergence of performance appraisal and management within the NHS and within health- and social-care regulation more generally. Importantly, in the context of fitness to practice and patient complaints, the relatively consistent administrative approach adopted by the GMC towards the handling of complaints in terms of the disposal pathway by which cases typically progress could be said to be demonstrative of a growing organisational reliance on codified risk-averse procedural rules to assist in the day-to-day processing of complaints. There is a very real danger here that this may over time undermine the value placed on the tacit dimensions of professional expertise within the broader professional community as rank-and-file practitioners in particular become ever more wary of the GMC and its associated bureaucratic machinery. This danger is perhaps even more pronounced now that the GMC is no longer under the direct control of medical elites in the form of the royal colleges and medical schools. Certainly, their influence in the operation of the GMC is still present, but the elected nature of the medical and non-medical members of the GMC means this now by and large operates by more indirect means than previously. While from the perspective of rank-and-file practitioners the changes under way in the GMC may well be looked at with some concern. Consequently, it can be argued that the concept of defensive medical practice is instructive here.

Defensive medicine occurs when diagnostic or therapeutic measures are used by a doctor as protection against possible accusations of negligence or underperformance, rather than because their patient really needs them (Summerton 1995). Studies show rank-and-file doctors are increasingly engaging in defensive medicine as a result of a rise in health-care managerialism, an increase in patient complaints and greater emphasis being placed on patient choice (Nettleton 2006). This reinforces the need to approach raw complaint data (and fitness to practice activity prior to the adjudication stage at least) with some caution. For a complaint can arise because of tension between a patient's sense of personal entitlement and strategic health-care planning and rationing as much as because of the action of an attending medical practitioner.

But what is most useful about the concept of defensive medicine is that it provides an illustrative example of how the reliance on codified and routinised frameworks to guide action within health-care systems can alter the behaviour of the wider professional community in unforeseen ways as practitioners seek to adjust to changing circumstance and avoid the possibility of punitive action being taken against them. One can certainly imagine this approach being adopted by some practitioners in relation to the revalidation arrangements discussed in Chap. 6. There is a real danger that the growth of risk-averse medical regulation may bring with it unintended negative consequences for patient care. Consequently, it is arguable that sociologists and other observers with an interest in medical regulation and the governance of professional forms of expertise need to carefully monitor the impact of GMC reform on practitioner behaviour in everyday clinical decision-making situations. This is a point which will be returned to in Chap. 8.

This brings us to another important point relating to the patient perspective and experience. This is concerned with recent developments regarding the role of the GMC in the hearing of complaints. As has already been discussed, the 2008 Health and Social Care Act established the Office of Health Professions Adjudicator (OHPA) to take over the role of the GMC in the adjudication of fitness to practice cases. Yet, in the summer of 2010, the UK government concluded that it was not persuaded of the need to introduce another regulatory body to take over the role of adjudicator in fitness to practice cases (Department of Health 2010). The key issue here, from the perspective of patients, is that it is arguable that the fact that a significant number of complaints do not make it past the initial triage and investigative stages raises legitimate questions about the GMC's gatekeeper role at each point in the decision-making and follow-up process. The issue here is not a lack of action being taken against a doctor. Rather it is the lack of rigorous, ongoing and publicly accountable third-party surveillance and appraisal of the reasons for a lack of action.

It certainly can be argued that the reforms made to the GMC have successfully turned it into a more autonomous regulatory body which is not as tied to the medical club in the form of the medical schools or royal colleges as it historically has been. However, here it must be remembered that medical elites do continue to exercise control over the GMC which does raise the legitimate concern that the influence of the medical profession on the adjudication process is such that the introduction of a completely independent party to judge appropriate punishment is arguably needed. Not least of all because the little independent research into the GMC handling of complaints which exists has in the past revealed the presence of judgemental bias (i.e. Allen 2000; Smith 2005). A recent small-scale independent review of a sample of GMC complaints found that *'articulate individuals who present their complaints clearly and in detail are more likely to have their cases taken up by the GMC'* (Hughes et al. 2007:15). Indeed, Hughes et al. (2007) point out that although their research was limited it is clear that GMC procedures need to be more sensitive to complainants needs, including their ability (or lack thereof) to effectively communicate their experience and concerns. The danger here is that complaints which should be taken forward by the GMC, or the local and national NHS complaint bodies it liaises with, may slip through the cracks and lead to a doctor being able to continue to practise. This state of affairs brings with it the very real danger of further potentially fatal risk to other patients. Indeed, although generally supportive of the GMC, the Council for Health Care Regulatory Excellence has stated in light of their recent audit of GMC operations that '*we consider that it [The GMC] needs to ensure that its decision makers have fully understood all the complainant's concerns, and that complainants feel that they are encouraged to submit a complaint*' (Council for Health Care Regulatory Excellence 2010b:28).

It is important for medical practitioners and regulatory bodies such as the GMC to remember that complaints are not made lightly or easily by individuals. Research shows that complainants know full well that a compliant can be seen as a hostile act (Mulcahy 2003). So they are often reluctant to come forward as well as frequently feel they must justify their actions by providing a highly personal and emotionally charged narrative concerning how they have been affected by what has happened (Nettleton 2006). On occasion, this narrative can stand in stark contrast to the tendency for health-care practitioners and their professional and regulatory organisations to focus more upon the technical aspects of treatment while often simultaneously discounting the ability of members of the public to fully understand why certain actions have been undertaken in particular situations instead of others (Stacey 2000). There is an inherent power imbalance in the relationship between the medical profession and general public. This is due in no small part to the esoteric specialist expertise that joins a doctor and patient together into a symbiotic relationship in the first place. This reinforces why independent third-party monitoring and appraisal of complaint decision-making processes is vitally important to the continued legitimacy of professional self-regulation as a governing strategy (Mulcahy 2003). Not least of all because exclusory in-house normative tendencies surrounding what are and what are not appropriate complaints can all too easily develop when dealing with sensitive value-laden high-risk situations, such as professional practice situations that seem to call into question the competence or ethical probity of a medical practitioner (Lloyd-Bostock and Hutter 2008).

Bearing this in mind, it is arguable that the descriptive statistics outlined in this chapter reinforce the need for further independent research into the GMC case management and hearing process to ensure recent reforms do not serve to underplay the legitimacy and value of the patient experience and perspective in all its multi-dimensional forms (Mulcahy 2003; Nettleton 2006). Such a research agenda is arguably just as important as conducting further research into the medical practitioner experience of GMC reform, particularly as there is a need to ensure a rounded picture is obtained of the impact of contemporary reforms in medical regulation on both doctors and patients. Indeed, it is arguably essential that sustained analysis of patient complaint data is undertaken to ensure that the protection of the public interest lies at the centre of medical regulation and the activity of the GMC.

Conclusion

This chapter has outlined the impact of the current regulatory reform agenda on the handling of complaints by the GMC. The descriptive statistical data shows that the GMC is more likely to receive a complaint about older doctors, doctors who qualified overseas and male doctors (although complaints against female doctors seem to be rising as a result of the changing gender make-up of the profession). Furthermore, the disposal pathway by which complaints subsequently move through the investigation and adjudication stages reinforces that the GMC does seem to be adopting a more punitive stance towards medical practitioners when their clinical competence and/or ethical probity is called into question. In previous chapters, it was highlighted that although some elements of medical knowledge are open to standardisation, routinisation and performance control, as an applied discipline medicine nevertheless possesses a strong tacit dimension which cannot be fully codified and subject to third-party appraisal. Furthermore, it was also pointed out that even when medical expertise is open to codification and performance management (usually by hospital management) peer appraisal by other medical practitioners nevertheless remains the best yardstick from which to judge actual clinical performance and ethical probity. Hence, the medical professions' de facto collective control over its specialist expertise, as well as by extension the performance criteria for admission and licensing credentials, leads it to possess a strong position from which to counter and/or co-opt mechanisms designed to performance manage medical work and its quality control. Nevertheless, what seems to have happened in relation to the handling of fitness to practice cases and GMC reform is that the regulatory state has acted to transform medical autonomy to bring it into line with the risk-adverse governing regimes of the risk society.

It needs to be acknowledged that the introduction of state-endorsed regulatory reform and quality assurance frameworks means that professional decision-making processes and regulatory regimes over the last decade have become increasingly trapped inside a seemingly ever-expanding web of governmental surveillance and performance management, with the result that doctor's traditional clinical freedoms and self-regulatory privileges are gradually being transformed by a seemingly ever-growing governmental need to make the delivery of esoteric medical expertise and complex health-care services more amenable to surveillance, calculation and performance management (Kuhlmann and Saks 2008). But this does not mean that the restratification thesis is no longer valid: the fact of the matter is that medical elites are vitally important to the governmental risk-management project. Indeed, it is this very issue which will be explore further in Chap. 8. In the meantime, the end-of-chapter self-study activities will help you to consolidate what you have learnt before moving on to study the contents of Chap. 8.

Self-Study Activity

1. Write a 1,000-word essay which critically evaluates the impact of contemporary changes in the gender, race and ethnicity make-up of the medical profession on GMC membership and how it handles complaints against doctors.
2. Produce a 15-min PowerPoint presentation which outlines the process by which the GMC handles complaints and critically evaluates if contemporary developments in the hearing of fitness to practice cases reinforce that the GMC is adopting a more punitive stance towards complaints.
3. Write a 100-word essay which critically evaluates if contemporary trends in the handling of fitness to practice cases reinforce the legitimacy of the restratification thesis.

References

Allen, I. (2000). *The handling of complaints by the GMC: A study of decision making and outcomes*. London: Policy Studies Institute.

Allsop, J. (2002). Regulation and the medical profession. In J. Allsop & M. Saks (Eds.), *Regulating the health professions* (pp. 113–134). London: Sage Publications.

Allsop, J. (2006). Regaining trust in medicine: Professional and state strategies. *Current Sociology, 54*(4), 621–636.

Beck, U. (1992). *Risk society: Towards a new modernity*. London: Sage Publications.

Bratland, S. Z., & Hunskår, S. (2006). Medico-legal assessments of complaints against general practitioners. *Tidsskrift for den Norske Lægeforening, 126*(1), 66–69.

Case, P. (2011). Putting public confidence first: Doctors, precautionary suspension and the general medical council. *Medical Law Review, 19*, 339–371.

Chamberlain, J. M. (2009). Portfolio-based appraisal: Superficial or useful? *British Journal of Hospital Medicine, 70*(11), 176–177.

Chamberlain, J. M. (2010a). Regulating the medical profession: From club governance to stakeholder regulation. *Sociology Compass, 4*(12), 1035–1042.

Chamberlain, J. M. (2010b). Performance appraisal for doctors: A case of paperwork compliance. *Sociological Research Online, 15*(1), 1–11. Available online: http://www.socresonline.org.uk/15/1/8.html. Accessed 1 Dec 2011.

Checkland, K., McDonald, R., & Harrison, S. (2007). Ticking boxes and changing the social world: Data collection and the new UK general practice contract. *Social Policy and Administration, 41*(7), 693–710.

Commission, H. (2007). *Investigation into outbreaks at Maidstone and Tunbridge wells NHS trust*. London: Healthcare Commission.

Council for Healthcare Regulatory Excellence. (2005). *Statistical summary*. London: Council for Healthcare Regulatory Excellence.
Council for Healthcare Regulatory Excellence. (2006). *Statistical summary*. London: Council for Healthcare Regulatory Excellence.
Council for Healthcare Regulatory Excellence. (2007). *Statistical summary*. London: Council for Healthcare Regulatory Excellence.
Council for Healthcare Regulatory Excellence. (2008). *Statistical summary*. London: Council for Healthcare Regulatory Excellence.
Council for Healthcare Regulatory Excellence. (2009). *Statistical summary*. London: Council for Healthcare Regulatory Excellence.
Council for Healthcare Regulatory Excellence. (2010a). *Statistical summary*. London: Council for Healthcare Regulatory Excellence.
Council for Healthcare Regulatory Excellence. (2010b). *Fitness to practise audit report: Audit of health professional regulatory bodies' initial decisions*. London: Council for Health Care Regulatory Excellence.
Davies, C. (2004). Regulating the health care workforce: Next steps for research. *Journal of Health Services Research & Policy, 9*, 55–61.
Deech, R. (2009). *Women doctors: Making a difference*. London: Department of Health.
Department of Health. (2000). *An organisation with a memory*. London: The Stationery Office.
Department of Health. (2009). *Health care regulation: Tackling concerns nationally*. London: Department of Health.
Department of Health. (2010). *Fitness to practice adjudication for health professionals*. London: Department of Health.
Department of Health. (2011). *Data on written complaints in the NHS 2010–2011*. London: DOH.
Dyer, C. (2010a). GMC strikes off GP who arranged later abortion for her daughter. *British Medical Journal, 340*, 762.
Dyer, C. (2010b). Cardiologist let his ambition interfere with patients best interests, GMC is told. *British Medical Journal, 340*, 3510.
Dyer, C. (2010c). Surgeon found guilty of dishonesty is suspended for six months. *British Medical Journal, 341*, 1136.
Elston, M. A. (2009). *Women and medicine: The future*. London: Royal College of Physicians.
Etheridge, L., Sturrock, A., Conlon, L., & Dacre, J. (2009). GMC tests of competence: The assessment of poorly performing doctors. *Clinical Risk, 15*, 15–17.
Francis, R. (2010). *Independent inquiry into care provided by mid Staffordshire NHS foundation trust January 2005–March 2009*. London: The Stationery Office.
Freidson, E. (2001). *Professionalism: The third logic*. Cambridge: Polity Press.
Giddens, A. (1990). *The consequences of modernity*. Cambridge: Polity.
Giddens, A. (1991). *Modernity and self-identity: Self and society in late modernity*. Cambridge: Polity.
Giddens, A. (1999). Risk and responsibility. *Law Review, 62*(1), 1–10.
GMC. (2000). *Fitness to practice annual statistics*. London: GMC.
GMC. (2001). *Fitness to practice annual statistics*. London: GMC.
GMC. (2002). *Fitness to practice annual statistics*. London: GMC.
GMC. (2003a). *Stakeholder research: Executive summary*. London: GMC.
GMC. (2003b). *Fitness to practice annual statistics*. London: GMC.
GMC. (2004a). *Guidance on GMC's fitness to practise rules*. London: GMC.
GMC. (2004b). *Annual review*. London: GMC.
GMC. (2004c). *Fitness to practice annual statistics*. London: GMC.
GMC. (2009). *Good medical practice*. London: GMC.
GMC. (2010). *Fitness to practice: Annual statistics*. London: GMC.
GMC. (2011). *The state of medical education and practice in the UK*. London: GMC.
Growth from Knowledge. (2011). *Research into fitness to practise referrals, a report for the GMC*. London: Growth from Knowledge.

Hughes, J., Locke, R., & Humphrey, C. (2007). *An independent audit of decisions in the investigation stage of the GMC's fitness to practise process*. London: King's College London.

Humphrey, C., Gulliford, M., Esmail A., & Cohen, D. (2009). *Clarifying the factors associated with progression of cases in the GMC's fitness to practise procedures: Full research report, ESRC End of Award Report, RES-153-25-0101*. Swindon: ESRC.

Irvine, D. (2003). *The doctors tale: Professionalism and the public trust*. London: Radcliffe Medical Press.

Kuhlmann, E., & Saks, M. (2008). *Rethinking professional governance: International directions in healthcare*. Bristol: Policy Press.

Lloyd-Bostock, S., & Hutter, B. (2008). Reforming regulation of the medical profession: The risks of risk based approaches health. *Risk and Society, 10*(1), 69–83.

McDonald, R., Harrison, S., & Checkland, K. (2008). Incentives and control in primary health care: Findings from English pay-for-performance case studies. *Journal of Health, Organisation and Management, 22*(1), 48–62.

Mulcahy, L. (2003). *Disputing doctors: The socio-legal dynamics of complaints about medical care*. Maidenhead: Open University Press.

National Clinical Assessment Service. (2011). *Concerns about professional practice and associations with age, gender, place of qualification and ethnicity: 2009/2010 data*. London: National Clinical Assessment Service.

Nettleton, S. (2006). *The sociology of health and illness*. London: Polity Press.

Office of National Statistics. (2001). *UK population ethnicity data*. London: Office of National Statistics.

Power, M. (1997). *The audit society: Rituals of verification*. Oxford: Oxford University Press.

Smith, J. (2005). *Shipman: Final report*. London: Department of Health.

Stacey, M. (1992). *Regulating British medicine*. London: Wiley.

Stacey, M. (2000). The general medical council and professional self-regulation. In D. Gladstone (Ed.), *Regulating doctors*. London: Institute for the Study of Civil Society.

Summerton, N. (1995). Positive and negative factors in defensive medicine: A questionnaire study of general practitioners. *British Medical Journal, 310*, 27–28.

Taragin, M. I., Wilczek, A. P., & Karns, E. (1992). Physician demographics and the risk of medical malpractice. *American Journal of Medicine, 93*, 537–542.

The Bristol Royal Infirmary Inquiry. (2001). *Learning from Bristol: The report of the public inquiry into children's heart surgery at the Bristol Royal Infirmary 1984–1995*. London: The Stationery Office.

Waring, J., Dixon-Woods, M., & Yeung, K. (2010). Modernizing medical regulation: Where are we now? *Journal of Health Organization and Management, 34*, 540–555.

West, P., Bending, M., & Chaplin, S. (2006). *A descriptive analysis of fitness to practise data for 2005 complaints*. York: University of York, York Health Economics Consortium.

Winyard, G. (2009). The future of female doctors. *British Medical Journal, 338*, 2223–2224.

World Health Organization. (2011). *World health statistics 2011*. Available at: http://www.who.int/whosis/whostat/2011/en/index.html. Accessed 1 Dec 2011.

Chapter 8
Epilogue: Sociology, Medical Governance and Citizenship

Abstract This chapter brings together the key themes outlined in preceding chapters into a clear concluding summary of the current state of the field in relation to the sociological study of medical regulation. In doing so, this chapter highlights the urgent need for sociologists to establish a programme of longitudinal research to track over the coming decade the impact of current reforms in medical regulation on both doctors and patients. It is argued that only by doing this will it be possible to more fully develop the restratification thesis outlined in previous chapters. The chapter also argues for the need to bear in mind broader ongoing changes in the nature of governance more generally, as the risk-saturated social conditions associated with high modernity continue to unfold around us. Here, the chapter discusses the need for sociologists to be aware of the inherent limitations of neo-liberal forms of governmentality and the types of subjectivity and models of good citizenship they promote. The chapter ends by inviting the reader to establish their own research agenda in light of the material and ideas discussed throughout this book.

Introduction

The previous chapters of this book have been concerned with introducing the reader to the sociological analysis of the medical profession and how it is regulated. In achieving this aim, current key developments in regulatory policy and practice have been outlined and critically discussed. As have the various conceptual frameworks sociologists use to explore the governance of medicine. Empirical data has been presented to help forge links between recent policy developments in medical regulation and trends in contemporary sociological theory pertaining to the study of medical governance. It has been highlighted how we now sit at the beginning of an important moment in the history of the governance of professional forms of expertise. Over the last decade, the regulatory state has intervened to transform medical autonomy through reforming medicine's regulatory institutions, arguably with the goal of bringing them in-line with the disciplinary imperatives of the risk society, with the result

J.M. Chamberlain, *The Sociology of Medical Regulation: An Introduction*,
DOI 10.1007/978-94-007-4896-5_8,

that medicine is increasingly dividing into more pronounced elite and rank-and-file segments as its shifts to a new governing regime. Bound up with this transformatory process has been the introduction of professional practice profiling using best-evidenced protocols, audit, as well as a mixture of individual and organisational standard setting and performance appraisal. This state of affairs will undoubtedly have a long-term effect on how practitioners approach their work in addition to how patients experience health-care delivery.

Arguably, the key task for sociologists over the coming decade is to empirically investigate the impact of current reforms on both doctors and patients, in no small part through monitoring the hearing of fitness to practice cases as well as reviewing newly introduced regulatory performance surveillance tools such as revalidation. True, under the new governing regime, one can expect medical elites themselves to engage in their own research project, but there will still be a need for independently conducted research. Yet, as was noted in the first chapter of this book, it will be vitally important to bear in mind while completing such a research task that current reform in medical governance takes place against the broader background of the fluxing social conditions associated with high modernity. Consequently, there is a need in this final chapter to bring together the themes discussed in this book into a clear concluding summary. Let us start this thematic ordering by beginning as we did in Chaps. 2 and 3 with the emergence of modern biomedicine.

Biomedicine, Peer Appraisal and Restratification

Chapter 2 traced the development of biomedicine and how modern medicine has emerged as a craft skill, combining both the fruits of the scientific method and tacit knowledge of experiential expertise, mastery of which can only be acquired through a protracted training programme. While the progressive and constantly developing nature of modern medical practice means that even when such formal training is completed there will be a need for subsequent periodic updating of a practitioner's knowledge and skills for as long as the doctor practises. As was discussed in Chaps. 3 and 4, it is medicine's prolonged regime of training and accreditation for its neophyte members, as well as the highly technical and specialised nature of modern medical practice, which protects it somewhat from external monitoring and evaluation. Certainly, sociologists of all persuasions need to bear in mind the truism that peer appraisal remains the core method by which the quality of medical training, practice and regulation can be assured in such a way as to minimise the threat of risk as far as is humanly possible (Freidson 2001).

It is often asserted by sociologists and medical historians that the pre-eminence of the modern medical profession lies in its scientific knowledge base, and in turn, this is linked to the historical development of pathological anatomy and the establishment of the hospital clinic as a site for the application of biomedicine. It is because of this that medical sociologists often use the terms 'biomedicine', 'medical model', or even 'biomedical model', as a shorthand ways of describing the dominant approach to the identification and treatment of illness and disease within western societies. As was discussed in Chap. 2, here the doctor is viewed as a mechanic who is treating a defective machine with presenting signs and symptoms being related through the application of medical science and the personalised clinical expertise of an individual practitioner, to underlying physical processes and abnormalities. Whatever its limitations, the underpinning machine view of the human body possessed by biomedicine has proven incredibly useful as a driving conceptual narrative from which medicine over the last 200 years has achieved remarkable success in combating illness and disease as well as relieving suffering and extending life. It is difficult to underestimate the positive influence modern medicine has had on improving the general quality of life of the average citizen in Western nation-states. Even the most ardent critic of entrenched medical power must acknowledge the beneficial role medicine continues to play within society.

It is also important to bear in mind the objective and rational basis of much of modern medical science, as was discussed in Chaps. 3 and 4. For this means it is open to standardisation, which in turn, at least since the rapid development of computer technology over the last 30 years, allows for the development of best-evidence practice protocols against which the performance of doctors can be evaluated, standardised and quality assured. Bearing in mind here the discussion in Chap. 1 of the key role medicine played in the dispersal of disciplinary power from the nineteenth century onwards, it is important to note that it is possible to some extent for the tables to be turned on the doctor-judge and for their own scrutinising norms for surveying and judging human behaviour to be used against them by non-medically trained individuals, including patients, politicians and the media. It is certainly the case that patient rights groups, politicians and health-care management, amongst others, have together over the last four decades increasingly sought to exploit the codifiable nature of much of everyday medical practice when seeking to curtail medical autonomy and make doctors more accountable for their actions. However, although scientific forms of expertise by their very nature are indeed open to codification, routinisation and standardisation, the everyday world of medical practice reinforces that practice of modern healing is as much an art as it is a science. As was discussed in Chaps. 4 and 5, medicine may well possess a formal scientific knowledge base, but there nevertheless is a high level of indetermination in the exercise of medical judgement and technique (Larson 1977). Hence, no matter how simplified the codification process is, the resulting protocols to guide action by and large often require the user to have completed some sort of formally overseen experiential training process, that is, if the protocols in question are to be applied in the least risky manner possible. In other words, in the final analysis it needs to be acknowledged that the highly specialised and yet experiential nature of medical expertise means some aspect of peer review and appraisal will remain a vital and essential part of quality assurance and everyday practice setting judgemental process.

As was discussed in Chaps. 3, 4 and 5 in some detail, the issue of the specialist nature of professional expertise, alongside the concurrent need for professionals to exercise discretion in their work, does create a buffer zone which arguably protects doctors from outsider surveillance and control (Freidson 2001). Yes, the emergence of evidence-based medicine has brought medical autonomy under the gaze of managerial performance imperatives based around concerns with cost and risk. The political rhetoric from both main political parties in the United Kingdom on the topic of the future of the National Health Service over the last three decades has used the growth of evidence-based medicine, medical audit and the clinical protocol to establish a regulatory guideline industry, as well as to bolster arguments for greater managerial control over medical work to improve service productivity, efficiency and safety. Furthermore, patients are increasingly unwilling to adopt the subservient position modern medicine has historically accorded them. Indeed, patients nowadays frequently see themselves as active health-care consumers who possess the right to make informed choices (Nettleton 2006).

Yet such considerations need to be balanced with the recognition that the nature of their knowledge demands that professionals must possess *'independence of judgment and freedom of action'* (Freidson 2001: 122). This is particularly the case if we as a society are to support and nurture the ongoing growth of professional practice communities and the quality of the services they provide. This is why although we may well be witnessing the increased use of standardised administrative procedures to performance manage medical work, we are also seeing the regulatory state accepting its limitations in relation to overseeing medical practice and the standard setting process surrounding it. A fact perhaps reinforced by the increasing reliance place on the royal colleges by the state as it seeks to introduce revalidation and reform the hearing of fitness to practice cases. As was discussed in some detail throughout this book, peer review remains *the* key tool by which the performance of doctors can be judged and in light of this it should come as no surprise that instead of going into decline the medical profession is undergoing a process of restratification into more pronounced elite and rank-and-file segments as it meets head on the growing need to reform how it manages its affairs. The former subjecting the latter to intensive performance surveillance and control in the face of external state-led pressure to transform medical regulation into a more appraisal orientated risk aware form. Yet in discussing recent reforms, such as revalidation and changes to the hearing of fitness to practice cases, Chaps. 6 and 7 together noted the limitations of the restratification thesis to fully capture the changes taking place. Indeed, it does seem that medical autonomy has been curtailed with the introduction of state-endorsed quality assurance frameworks, such as clinical governance. While it is undoubtedly the case that professional decision-making processes and regulatory quality assurance regimes over the last two decades have become increasingly trapped inside a seemingly ever-expanding web of governmental surveillance and performance management (Lloyd-Bostock and Hutter 2008).

For some, this state of affairs raises the question of the continued legitimacy of the restratification thesis as descriptive let alone explanatory analytical tool. But to my mind, I think it is important to acknowledge that we are sitting on the cusp of far-reaching changes in medical governance and so we as yet do not possess the necessary empirical data from which to fully analyse their impact. The discussion of revalidation in Chap. 6 and the hearing of fitness to practice cases in Chap. 7 recognised the need for sociologists to engage in a sustained programme of research to generate much needed empirical data from which to draw conclusions about the impact of current regulatory reforms on doctors and patients. Not least of all because it is only through the rigorous collection of such empirical data that the validity of the restratification thesis can be ascertained or an alternative theory of events developed. Yet there is also a further reason for engaging in a programme of sustained empirical research. As has also been pointed out at various points in the previous chapters, there is a need to recognise the role being played in shaping current events in the field of medicine by the fundamentally transformative nature of neo-liberal forms of governmentality and how these seek to embed disciplinary control within the population via the practice of a certain type of good governance which valorises the enterprise self. It is to this theme this chapter will now turn.

Good Governance and Good Citizenship in a Risky Age

> [Under liberal mentalities of rule] a person's relation to all his or her activities, and indeed his or her self, is...given the ethos and structure of the enterprise form.
>
> Rose (1999: 138)

If Chap. 1 highlighted the historical role medicine played in legitimising a new form of disciplinary surveillance and control from the nineteenth century onwards, then Chap. 2 reinforced the success of this project through noting how the dispersal of discipline throughout all aspects of society has intensified over the last century via the medicalisation of the everyday life-world. It is because of this state of affairs that the analysis of medicine and risk is indelibly linked to a core disciplinary concern within sociology with the nature of good governance and good citizenship (Rose 1999). For underlying recent reforms in medical governance is a more fundamental shift in the conditions under which good governance and good citizenship can be practised as a result of the economic and political re-emergence of liberalism as societies globally shift into the age of risk (Rose and Miller 1992; Mythen 2004). As was discussed in previous chapters, for risk theorists a key defining feature of modern society is that there has been *'a social impetus towards individualisation of unprecedented scale and dynamism...[which]...forces people – for the sake of their survival – to make themselves the centre of their own life plans and conduct'* (Beck and Beck-Gernsheim 2002: 31). It is argued that as capitalist-industrial society gives way under the tripartite forces of technology, consumerism and globalisation, there has been a categorical shift in the nature of social structures, and more importantly, the relationship between the individual and society. Furthermore, as working conditions change, and the technology and communication revolutions continue at pace, more than ever before individuals are required to make life-changing decisions concerning education, work, self-identify and personal relationships, in a world where traditional beliefs about social class, gender and the family are being overturned (Lupton 2011). This state of affairs leads to a concern with risk management entering centre stage within society's institutional governing apparatus, as well as individual subject-citizen's personal decision-making process, a situation that in turn arguably leads to a certain fundamental element of mistrust entering into the expert-public relationship. In short, we no longer know who we can trust as all previous sources of authority no longer seem valid.

The previous chapters have discussed in some detail how contemporary reforms in medical regulation can be viewed as an attempt to rebuild trust between the profession and public as the problem of medical risk, in terms of medical malpractice, negligence and accident, is met head on by the regulatory state. Osborne (1993) discusses how since the re-emergence of liberalism there has been a gradual reformulation of health-care policy and practice, so that 'the field of medicine' as he calls it, to a greater degree than ever before, is simultaneously both governed and self-governing. For Osborne and many other commentators, a key part of this process is the subjection of the activities of medical practitioners to an additional layer of management and new formal calculative regimes, both by their peers and by external groups such as health service management (Rose and Miller 1992). These calculative regimes include performance indicators, competency frameworks and indicative budget targets (Rose 1993). This process began with the 1979 conservative administration which possessed a firm neo-liberal commitment to rolling back the state and introducing free market philosophies within the public and private spheres (Dean 1999). Furthermore, the New Labour government of the 1990s and first part of the new millennium arguably continued this process, while it seems both present and future governments are forced to continue in a similar vein due to the broader economic environment.

Neo-liberal philosophies emphasise the entrepreneurial individual, endowed with freedom and autonomy, and a self-reliant ability to care for herself, who furthermore is driven by the desire to optimise the worth of her own existence. This conception of the social actor as a free enterprising self is the core mechanism by which neo-liberal mentalities of rule seek to tap into the self-regulatory capabilities of the individual so they can be entwined with the key objectives of governance – the security, health, wealth and happiness of the general population (Barry et al. 1996). It certainly can be argued that the introduction of performance appraisal tools such as revalidation for doctors is just one more example of the internationally recognised trend that, like many other professionals, doctors are becoming subject to a seemingly ever-increasing number of formal calculative regimes that seek to performance manage their work practices in order to better economise and risk-manage occupational tasks in the face of a shift towards a neo-liberal model of governance (Coburn and Willis 2000; Checkland et al. 2007; McDonald et al. 2008). Power (1997), like many others, emphasises the enormous impact of the contemporary trend in all spheres of western societies towards Audit in all its guises – with its economic concern with transparent accountability and standardisation – particularly for judging the activities of experts. This, as was discussed in detail in Chap. 5, is bound up with the re-emergence of liberalism as an economic and political philosophy (Rose 1996). Against this background Townley (1993a, b, 1997, 1999), Newton and Findley (1996) and Rose (1999) all suggest performance appraisal in all its various guises is a distinctive form of neo-liberal governmentality: a system of control which utilises surveillance and rationality to turn the object of its gaze into a calculable and administrable subject open to control and risk management (Foucault 1991).

This Foucauldian interpretation of appraisal holds that it acts as an information panopticon that operates through the use of two key panoptic disciplinary mechanisms – normalisation and hierarchy (Zuboff 1988). Normalisation, or normalising judgements, involves comparing, differentiating and homogenising in relation to assumed norms or standards of what is proper, reasonable, desirable and efficient (Foucault 1979). Appraisal possesses normalising judgements due to its focus upon establishing behavioural norms in the form of 'on-the-job' task standards from which to judge individual performance. Hierarchy involves a process of judging, ranking and rating an individual without in turn being judged (Foucault 1979). This reinforces that no matter how much its advocates hold it is user-centred and developmental performance appraisal is nevertheless a punitive disciplinary tool concerned with identifying areas of under-performance and correcting them (Fletcher 1997). Yet appraisal is not a straightforward punitive disciplinary tool, concerned with identifying and correcting poor performance 'from the outside' (Rose 1996). Indeed, this Foucauldian interpretation of appraisal holds that it may seek to promote and reward certain behaviours and rectify others, but it recognises that it nevertheless more often than not does so by operating using a more subtle and invasive form of soft power (Rose 1999). Certainly, within medicine, appraisal seeks to work on the subjectivity of appraisees at a distance through requiring they engage in self-surveillance of their clinical performance as if it were a normal and everyday practice as a result of the availability of best-evidenced clinical guidelines and protocols (Sheaff et al. 2003). For example, for annual NHS appraisal and more lately revalidation purposes, consultants and general practitioners must keep a portfolio of their continuing professional development needs and fitness to practise which contains personalised information relating to prescribing patterns, the outcomes of case note analysis, the results of clinical audit, as well as patient complaint case outcomes and surgical operation success rates (Black 2002). They must use this information to help identify and publicly record areas of developmental need in relation to best-practice performance frameworks, guidelines and protocols (Bruce 2007). Furthermore, they must subsequently record activities and achievements that demonstrate they are proactively meeting their self-identified learning goals, which will subsequently be subject to formal peer review, to prove they are willing as a matter of good professionalism to admit to areas of poor performance and learn from them (Irvine 2003). It will perhaps come as no great surprise then to learn then that individuals who advocate performance appraisal within medicine argue that it simply formalises what should already be a normal and natural part of a doctor's day-to-day self-monitoring of their clinical performance (see Snadden and Thomas 1998; Davis et al. 2001; Wilkinson et al. 2002).

Furthermore, as has been argued in this book, this invasive soft power style of governance which is transforming the operation of medical governance is in fact a generic feature of contemporary forms of rule. Indeed, they are bound up with the promotion of what can be termed 'good citizenship' (Rose 1999). Certainly, even if we reject the risk society thesis, we can agree that there has been a profound shift in *'the nature of the present'* (Rose 1992: 161) and the way *'[we] come to recognise ourselves and act upon ourselves as certain kinds of subject'* (Rose 1992: 16). Due in no small part to the re-emergence of liberalism and the growing ascendancy of the concept of the enterprise self throughout all spheres of modern social life (Gordon 1996). For example, Burchell (1996) argues that neo-liberalisms dual advocacy of the self-regulating free individual and the free market has led to *'the generalisation of an "enterprise form" to all forms of conduct'* (Burchell 1996: 28). Enterprise – with its focus upon energy, drive, initiative, self-reliance and personal responsibility – has assumed a near-hegemonic position in the construction of individual identities and the government of organisational and everyday life. Enterprise has assumed *'an ontological priority'* (du Guy 1996: 181). Consequently, as Burchell (1993: 275) notes, *'one might want to say that the generalization of an "enterprise form" to all forms of conduct – to the conduct of organisations hitherto seen as being non-economic, to the conduct of government, and to the conduct of individuals themselves – constitutes the essential characteristic of this style of government: the promotion of an enterprise culture'.*

Such considerations remind us that changes in how expertise operates are directed towards the object of good governance – the population in general and the individual subject-citizen in particular – as much as they are experts themselves (Rose 1999). For changes in how participatory citizenship is practised are bound up with shifts in the conditions under good governance operates. In terms of Berlin's (1969) famous dichotomy of 'positive' and 'negative' liberty, although liberal mentalities of rule may appear at first to promote negative liberty (i.e. the personal freedom of the individual-subject to decide who they are and discover what they want to be), in reality they promote positive liberty (i.e. a view of who and what a citizen-subject is and should be). It certainly can be argued that a key facet of advanced liberal society is its central concern with disciplining the population without recourse to direct or oppressive intervention. Yet liberal mentalities of rule seek to promote good citizenship by discursively constructing and promoting subjective positions for subject-citizens to occupy in relation to the forms of the enterprise self. Typically, this is associated with a bundle of characteristics such as energy, resilience, initiative, ambition, calculation, self-sufficiency and personal responsibility (Rose 1996). For the world of enterprise valorises the autonomous, productive, self-regulating individual, who is following their own path to self-realisation, and so it requires all society's citizens to *'come to identify themselves and conceive of their interests in terms of these...words and images'* (du Guy 1996: 53).

A key consequence of this state of affairs is that failure to achieve the goal of self-fulfilment is not associated with the possession of a false idea of what it means to be human. Nor is it that individuals do not possess an essential core self which is the real and true them for all eternity. Rather, such failure is deemed to be the result of poor choices, a lack of education or the dependency culture created by the welfare state. It is the result of learned helplessness, which in itself can be resolved with *'programmes of empowerment to enable [the individual] to assume their rightful place as self-actualizing and demanding subjects of an "advanced" liberal democracy'* (Rose 1996: 60). The sociological analysis of medicine, I would argue, needs to focus upon this point as it considers the type of citizen and forms of subjectivity promoted and sustained by the governing regimes of the risk society. For under the guise of advocating minimal forms of government as the natural way of things, liberal mentalities of rule run the risk of promoting a highly limiting view of what it is to be a human being, let alone a good citizen, within today's increasingly complex social world. Because in arguing that individual's sense of self is now arguably more than ever before a product of her own making, such conceptions stay wielded to the idea of the subject as an autonomous actor possessing a coherent core self (Elliott 2001). That is, they can be said to emphasise a *'positivist ego psychology, which is hostile to any notion that the self is complexly structured and differentiated'* (Peterson 1997: 190). This leads to a state of affairs where *'modern individuals are not merely "free to choose", but obliged to be free, to understand and enact their lives in terms of choice'* (Rose 1999: 87). Not least of all because for neo-liberalism it is through the exercise of personal choice that the self is held to be realisable to both oneself and others. In contrast to such possessive individualism, following Foucault (1991), governmentality theorists firmly historicise their conception of the individual by discursively locating it within the history of western thought through critiquing the development of neo-liberalism as being tied up with a post-enlightenment conception of a rationally autonomous subject (Peters 2001). Instead, they advocate an alternative viewpoint whereby individual subjectivities are neither fixed nor stable, but rather are constituted in and through a spiral of power-knowledge discourses – generated by political objectives, institutional regimes and expert disciplines – whose primary aim is to produce governable individuals (Deleuze 1988). Such a conception arguably better fits the fluxing social circumstances associated with the individual's everyday experience of modern life today. Furthermore, this alternative view of human nature and agency may well offer an avenue for achieving positive social change. However, such considerations are beyond the scope of this text.

Conclusion

In conclusion, given the events discussed throughout this book, I would argue that social scientists interested in the study of medical regulation need to remember that current developments in the governance of doctors must be analysed within the broader socio-economic and political context that influenced their development. In other words, medicine's regulatory arrangements are not solely the result of the medical professions possession of disciplinary expertise or an ethical patient-centred orientation. Recent reforms in medical governance are to no small measure bound up with a broader ongoing shift in how good governance is conceptualised and operationalised under neo-liberal mentalities of rule as the regulatory state seeks to utilise a mixture of both hard and soft forms of power to promote a certain type of citizen-subject congruent with the enterprise form. As was touched upon in Chap. 1, this governing goal is sought within the risk-saturated conditions associated with high modernity offers an opportunity to examine how forms of individual resistance emerge due to the technological and communication advances that characterise this particular period in history. A key question to answer here is if such developments help to transform existing elitist and sometimes discriminatory social hierarchies, while at a practical level I think it clearly important for social scientists interested in the study of medical regulation to engage in a dedicated longitudinal research programme concerned with the implementation of new regulatory quality assurance tools, such as revalidation. Not least of all because this is a research area where social scientists could use their own expertise to help medical elites strike a balance between protecting a necessary degree of professional autonomy and ensuring medical work remains open to an element of independent surveillance and control in order to help protect the general public from poorly performing medical practitioners. After all, medicine is not the only profession that possesses a strong public service ideal. Indeed, before putting this book to one side, the reader might like to complete the following final self-study activity.

Self-Study Activity

1. Draw together an action plan of how you would conduct an empirical research project analysing the impact of contemporary changes in medical regulation, that is, revalidation and the hearing of fitness to practice cases, on both doctors and patients. Outline in some detail the research questions you will ask, the research method you will use to answer them, as well as how you will go about getting people to participate in your study. Do not forget to also discuss how you think your research questions will help you to critically contribute to the argument that the medical profession is currently undergoing a process of restratification.

References

Barry, A., Osborne, T., & Rose, N. (1996). *Foucault and political reason*. London: University College London Press.

Beck, U., & Beck-Gernsheim, E. (2002). *Individualization: Institutionalized individualism and its social and political consequences*. London: Sage Publications.

Berlin, I. (1969). Two concepts of liberty. In I. Berlin (2002) *Four essays on liberty* (pp. 14–35). Oxford: Oxford University Press.

Black, D. (2002). *Consultant appraisal GMC news supplement on appraisal and revalidation 1–4*. London: The GMC.

Bruce, D. A. (2007). Regulation of doctors. *British Medical Journal, 334*, 436–437.

Burchell, G. (1993). Liberal government and techniques of the self. *Economy and Society, 22*(3), 267–282.

Burchell, G. (1996). Liberal government and techniques of the self. In A. Barry, T. Osborne, & N. Rose (Eds.), *Foucault and political reason* (pp. 133–167). London: University College London Press.

Checkland, K., McDonald, R., & Harrison, S. (2007). Ticking boxes and changing the social world: Data collection and the new UK general practice contract. *Social Policy and Administration, 41*(7), 693–710.
Coburn, D., & Willis, E. (2000). The medical profession: Knowledge, power and autonomy. In G. L. Albrecht, R. Fitzpatrick, & S. C. Scrimshaw (Eds.), *The handbook of social science and medicine* (pp. 257–277). London: Sage Publications.
Davis, M. H., Friedman, B. D., Harden, R. M., Howie, P., Ker, C., McGhee, M. J., Pipard, M. J., & Snadden, D. (2001). Portfolio assessment in medical students' final examinations. *Medical Teacher, 23*(4), 231–250.
Dean, M. (1999). *Governmentality: Power and rule in modern society*. London: Sage Publications.
Deleuze, G. (1988). *Foucault*. Minneapolis: University of Minnesota Press.
du Guy, P. (1996). *Consumption and identity at work*. London: Sage Publications.
Elliott, P. (2001). *Concepts of the self*. Cambridge: Polity Press.
Fletcher, C. (1997). *Appraisal: Routes to improved performance*. London: Short Run Press.
Foucault, M. (1979). *Discipline and punish: The birth of the prison*. London: Allen Lane.
Foucault, M. (1991). Governmentality. In G. Burchell, C. Gordon, & P. Miller (Eds.), *The Foucault effect: Studies in governmentality* (pp. 11–27). London: Harvester Wheatsheaf.
Freidson, E. (2001). *Professionalism: The third logic*. Cambridge: Polity Press.
Gordon, C. (1996). Foucault in Britain. In A. Barry, T. Osborne, & N. Rose (Eds.), *Foucault and political reason* (pp. 167–184). London: University College London Press.
Irvine, D. (2003). *The doctors tale: Professionalism and the public trust*. London: Radcliffe Medical Press.
Larson, M. S. (1977). *The rise of professionalism: A sociological analysis*. California: University of California Press.
Lloyd-Bostock, S., & Hutter, B. (2008). Reforming regulation of the medical profession: The risks of risk based approaches. *Health, Risk and Society, 10*(1), 69–83.
Lupton, D. (2011). *Medicine as culture* (3rd ed.). London: Sage Publications.
McDonald, R., Harrison, S., & Checkland, K. (2008). Incentives and control in primary health care: Findings from English pay-for-performance case studies. *Journal of Health, Organisation and Management, 22*(1), 48–62.
Mythen, G. (2004). *Ulrich beck: A critical introduction to the risk society*. London: Pluto.
Nettleton, S. (2006). *The sociology of health and illness*. London: Polity Press.
Newton, T., & Findlay, P. (1996). Playing god? The performance of appraisal. *Human Resource Management, 6*(3), 42–58.
Osborne, T. (1993). On Liberalism, neo-liberalism and the liberal profession of medicine. *Economy and Society, 22*(3), 345–56.
Peters, M. (2001). *Poststructuralist, Marxism and neo-liberalism: Between theory and practice*. Lanham: Rowman and Littlefield.
Peterson, A. (1997). Risk, governance and the new public health. In A. Peterson & R. Bunton (Eds.), *Foucault health and medicine* (pp. 3–21). London: Routledge.
Power, M. (1997). *The audit society*. Oxford: Oxford University Press.
Rose, N. (1992). Governing the enterprise self. In P. Heelas & P. Morris (Eds.), *The values of the enterprise culture* (pp. 46–65). London: Routledge.
Rose, N. (1993). Government, authority and expertise in advanced liberalism. *Economy and Society, 22*(3), 283–299.
Rose, N. (1996). Governing advanced liberal democracies. In A. Barry, T. Osborne, & N. Rose (Eds.), *Foucault and political reason* (pp. 102–123). London: University College London Press.
Rose, N. (1999). *Powers of freedom: Reframing political thought*. Cambridge: Cambridge University Press.
Rose, N., & Miller, P. (1992). Political power beyond the state: Problematics of government. *The British Journal of Sociology, 43*(2), 173–205.

Sheaff, R., Rogers, A., Pickard, S., Marshall, M., Campbell, S., Sibbald, B., Halliwell, S., & Roland, M. (2003). A subtle governance: 'Soft' medical leadership in English primary care. *Sociology of Health & Illness, 25*(5), 408–428.
Snadden, D., & Thomas, M. (1998). The use of portfolio learning in medical education. *Medical Teacher, 20*(3), 244–265.
Townley, B. (1993a). Foucault, power/knowledge, and its relevance for human resource management. *Academy of Management Review, 18*(3), 518–545.
Townley, B. (1993b). Performance appraisal and the emergence of human resource management. *Journal of Management Studies, 30*(2), 221–238.
Townley, R. (1997). The institutional logic of performance appraisal. *Organization Studies, 18*(2), 261–285.
Townley, R. (1999). Practical reason and performance appraisal. *Journal of Management Studies, 36*(3), 287–306.
Wilkinson, T. J., Challis, M., Hobma, S. O., Newble, D. I., Parbossingh, J. T., Sibbald, R. G., & Wakefield, R. (2002). The use of portfolios for assessment of the competence and performance of doctors in practice. *Medical Education, 36*(1), 23–46.
Zuboff, S. (1988). *In the age of the smart machine*. Oxford: Heinemann.

MIX
Papier aus verantwortungsvollen Quellen
Paper from responsible sources
FSC® C105338

If you have any concerns about our products, you can contact us on
ProductSafety@springernature.com

In case Publisher is established outside the EU, the EU authorized representative is:
Springer Nature Customer Service Center GmbH
Europaplatz 3, 69115 Heidelberg, Germany

Printed by Libri Plureos GmbH
in Hamburg, Germany